P9-CMX-206

THE COMPLETE SOLAR HOUSE

THE COMPLETE SOLAR HOUSE

Bruce Cassiday

Illustrated with photographs and diagrams

DODD, MEAD & COMPANY
NEW YORK

ALSO BY BRUCE CASSIDAY

HOW TO CHOOSE YOUR VACATION HOUSE

Printed in the United States of America

3 4 5 6 7 8 9 10

Library of Congress Cataloging in Publication Data

Cassiday, Bruce.
The complete solar house.

Includes index.
1. Solar houses. I. Title.
TH7413.C37 697'.78 77-21461
ISBN 0-396-07493-6

Acknowledgments

I would like to express my personal thanks to the following people and organizations for their help in making this book possible:

Dave Rogoway, American Plywood Association; Robert C. Carmody, Copper Development Association; Pamela Allsebrook, California Redwood Association; Douglas A. Cornell, Aluminum Association; John Baer, Mineral Wool Association; Elizabeth M. Rich, Carpet and Rug Institute and Society of the Plastics Industry; National Solar Heating & Cooling Information Center; Energy Research and Development Administration; National Aeronautics & Space Administration; Department of Housing and Urban Development; University of Maine, Orono, Me.; Floyd Perron, Sunworks, Inc.; Harold R. Hay, Skytherm Process & Engineering; Tom Minnasion, Kalwall Corporation; Jesse J. Savell, Jr., Savell System; Robert J. Schlesinger, Rho Sigma; Roy Williamson, Solaron Corporation; Doug Taff, Garden Way Laboratories; Jerry Arkebauer, Owens-Illinois; E. H. Michaelsen, Phelps Dodge; Elizabeth Kilkenny, Westinghouse Electric Corporation; Fitzroy Newsum, Martin Marietta Aerospace; Joe Dawson, Grumman Corporation; F. James Carr, Tyco Laboratories, Inc.; Robert K. Spear, Arkla Industries, Inc.; Edmund Scientific Company; Reynolds Aluminum; Tranter, Inc.; Fred Rice Productions; Solar Systems, Inc.; Revere Copper & Brass, Inc.; Sol-Heet Sun-du Company; Fafco, Inc.

Contents

THE COMPLETE SOLAR HOUSE

1

What Solar Energy Is

The use of solar energy to heat homes is nothing new. When man built the first translucent window in his house to let in the sun's rays he was utilizing solar energy for the purpose of space heating.

But when winter came, and brought cloudy skies and storms, he found the sun's rays inadequate to the task. He turned first to wood fuel to provide him heat on demand, and then eventually turned to coal and oil and other fossil fuels, called fossil because they are in effect the remains of animal life buried beneath the earth's surface for millions of years.

In fact, fossil fuel—coal, petroleum derivatives, and natural gas—has been used since the beginning of the Industrial Revolution to provide heat for industry's machines. It is inexpensive, easy to procure from the earth, easy to move from one place to another, and quick to produce heat.

It is so productive that it is used today for almost 95.9 percent of the production of energy in the United States. Its basic drawback is its exhaustibility. The stock of fossil fuel on the planet earth is finite. In fact, its end is in sight. If its use escalates much in the near future, the entire stock of fossil fuel will be gone not in hundreds of years, but in decades.

Moreover, the cost of fossil fuel has become a potential weapon in an economic battle that is being waged between world power blocs. For that reason also, fossil fuel may soon be phased out.

Nuclear energy has possible destructive ancillary disadvantages in spite of its energy-producing potential. It too is exhaustible, with limitations to the amount of nuclear ore lodged in the earth's crust. Although the stockpile of nuclear energy is larger than that of fossil fuel, it too will not last forever. Today, only about .3 percent of the energy in the United States is produced by nuclear generators.

The rest of it not produced by fossil fuel—3.8 percent—is produced by hydroelectric generators operated on rivers and dams by the flow of water. Although hydroelectric generators are not generally thought of as examples of solar-energy machines, the flow of water that runs them is definitely caused by precipitation—an indirect result of the sun's heat.

HOW MUCH FUEL IS LEFT?

To deal with energy figures with so many zeros, scientists have established a unit called a Q. The Q represents the burning of about 39 billion tons of coal, 172 billion barrels of oil, or 968.9 trillion cubic feet of natural gas.

One Q is actually a quintillion BTUs, a quintillion being a billion billion, or the figure 1 with eighteen zeros after it. That figure is also written 1×10^{18}.

One BTU is one British Thermal Unit, the amount of heat needed to raise one pound of water one degree Fahrenheit. To give some idea of the amount of heat entailed, the average home furnace is rated at a maximum of 100,000 BTUs. It takes about 152 BTUs to make a pint of water boil—if the water's tap temperature is about 60° F.

The world's population burns up about one-tenth of a Q of fossil fuel per year. As of now, the earth has about 25 Q of fossil fuel reserves left, according to some estimates. Other estimates say that the reserve is ten times that much. No one really knows for sure.

As for nuclear reserves, the figure is put at about 575 Q. The total of those two figures, using 25 Q for fossil fuel, is around 600 Q.

And that doesn't look too bad, with a year's figure of total fuel

WESTINGHOUSE ELECTRIC CORPORATION
In the future, solar collectors like these will become commonplace as man fights to bridge the energy gap.

consumption set at about one-tenth of a Q, and a balance of much less through nuclear power.

However, that doesn't take into account the escalation of energy usage in recent years. For example, mankind by 1850 had used up only an estimated 9 Q of fossil fuel energy in his entire tour of the universe. From 1850 to 1950, however, almost 4 Q were consumed, or almost half that much. Predictions of energy consumption between 1950 and 2050 add up to 487 Q!

A few years after 2050, according to these predictions, there will be no more of this energy available.

With finite limits on fossil fuel and nuclear fuel so near exhaustion, it is obvious that man must turn to other sources to locate supplies of energy for the future.

Of the three kinds of energy used by man to run his industrial machines and to make his surroundings comfortable enough to live in, both fossil fuels and nuclear fuels have been exploited effectively. A third fuel source—solar energy—has not.

WHERE THE SUN GETS ITS ENERGY

Solar energy is produced on the sun's surface by the nuclear process of fusion. It is the result of the fusion of individual hydrogen atoms under the intense pressure of the sun's gravity and magnetism. In the fusion process, the hydrogen atoms become helium atoms. As that happens, a tremendous amount of heat is released—an estimated 10,000° Fahrenheit on the sun's surface.

Only a small percentage of this energy reaches the earth after a 93,000,000-mile journey through space. At an altitude of 100 miles above the earth's surface, one square foot of atmosphere receives about 430 BTUs per hour in solar energy.

In its last 100 miles of travel through the earth's atmosphere, the energy in that square foot of area is reduced at the earth's surface to about 300 BTUs. At northern and southern latitudes, the amount is less. Cloudy and hazy conditions further reduce the figure to as low as 50 BTUs. Seasonal conditions also affect the specific amount.

For example, the continental United States contains areas which range in yearly sunshine from about 2,200 hours in Maine, upper Michigan, and western Washington to about 4,000 hours in the Mohave Desert of California and Arizona.

Solar radiation received from the East Coast to the Rocky Mountains in the winter may be as little as one-fifth to one-third the amount received in the summer.

Of the energy that does reach the earth, about 30 percent is reflected back into space. Of the remainder, almost 47 percent is absorbed by the atmosphere, the land, and the oceans. This energy is stored as heat. Some is reradiated back into space, preventing accumulations of heat from reaching amounts that would be hazardous or lethal to plant and animal life.

About 23 percent of the sun's energy that reaches the earth becomes involved in a cycle of evaporation and precipitation before it once again reradiates out into space.

Less than 1 percent produces winds, oceanic circulation, and atmospheric disturbances.

Of the energy received by the earth, less than .03 percent is utilized by plants in the process of photosynthesis, producing the growth of plants and the cultivation of food that supports all animal life.

In the natural process of radiation and reradiation, enough heat passes through the atmosphere to take care of all the natural functions mentioned above with still enough left over to provide man with all his energy needs.

HOW TO PUT THE SUN TO WORK

Solar energy, experts agree, is an essentially inexhaustible source potentially capable of meeting a significant portion of the nation's future energy needs with a minimum of adverse environmental consequences. Indications are that solar energy is the most promising of the unconventional energy sources used by man.

For example, figures show that 2 trillion kilowatt hours of electricity are consumed in an average year in the United States. The amount of solar energy falling only upon American deserts averages 186 kilowatt hours per square foot. A total of 2 trillion kilowatt hours of electrical energy can be produced by capturing the amount of solar radiation falling on only 400 square miles of desert land. In fact, even assuming that the efficiency of conversion to electric power is only 5 percent, the total electrical energy used by the country can be captured on 8,000 square miles of desert—an area less than 10 percent of our total desertland.

At least 9 quadrillion kilowatt hours of solar energy fall each year on the continental United States, a sum equivalent to the energy available from 1.15 trillion tons of coal or 4.25 trillion barrels of oil.

Harnessing only one part of the solar energy for every 137,000 parts that are available will save the equivalent of 730 million barrels of oil yearly.

Estimates show that the country can become as solar efficient as that in twenty-five years. It is just a matter of getting efficient collector systems developed to grasp and use this free energy.

Two ways of creating electrical energy out of solar energy are known today. One of them involves the absorption of solar energy as heat and the subsequent conversion of heat into steam to turn turbines for the generation of electrical energy. The second involves the development of solar energy cells, or batteries, which can directly convert solar energy into electrical energy. These cells work

PHELPS DODGE BRASS COMPANY
Scientists at Pittsburgh research center monitor solar absorption not only with south-facing panels, but panels with eastern and western orientations as well.

the same way as rechargeable batteries, except that solar energy charges them instead of electric power. (See Chapter Fifteen)

IT'S FREE FOR THE TAKING

Solar energy is free for everyone. It is clean, producing no nuclear radiation nor dangerous fallout, no smoke nor air pollution. And yet man at the present time is not making use of solar energy to run the machines of industry. He is not even using solar energy to warm himself and make his shelters comfortable to live in.

It makes no sense, seemingly, and yet because of economic factors it *does* make sense.

Furnaces and machines fueled by petrochemicals and operated by nuclear piles are cheaper to build than machines that operate from solar energy. During the years between the beginning of the Industrial Revolution and the early 1970s, coal, oil, and natural gas were cheap and easy to obtain. Machines run by fossil fuel simply cost less to operate than those run by solar energy.

Some of the first heat machines invented in the eighteenth century, in fact, were operated by solar energy. The use of coal produced a fiercer and more concentrated fire, and provided the user with better service. Fossil fuel won out over solar energy.

During the process of developing the machines of the Industrial Revolution, scientists discovered enough about solar energy to make such machinery competitive in every way with fossil fuel machines —except economically.

It is from these early solar energy machines, called "sun heat" machines, that today's technology borrows techniques and methodology.

Solar-heated and -cooled public institutions like this school are isolated instances today. Harnessing only 1/137,000th of available solar energy will save 730 million barrels of oil yearly.

WESTINGHOUSE ELECTRIC CORPORATION

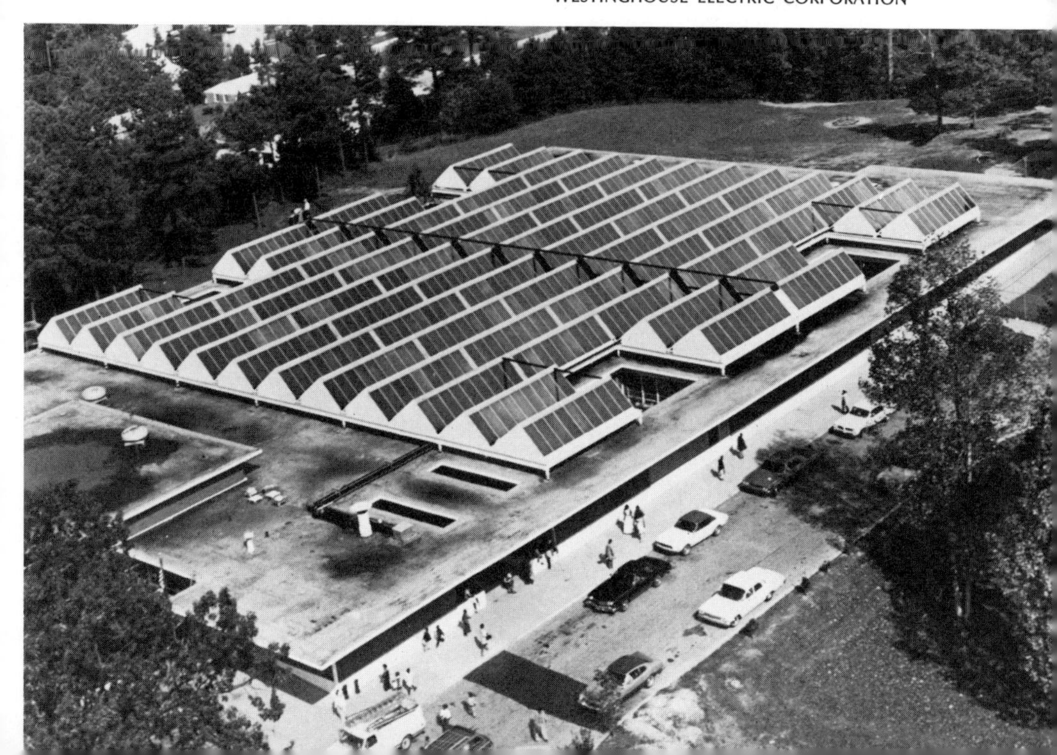

EARLY SOLAR ENERGY USAGE

Pure solar energy has been used by farmers for thousands of years to grow crops. Plant growth involves photosynthesis—the capturing of solar energy to promote growth.

Solar energy indirectly affects all weather, causing changes that include rain and wind. Moving air has been harnessed by man to push his ships from one place to another, and to turn his windmills to pump water and grind wheat.

It is solar energy that also affects all climate by producing the cycle of evaporation and condensation that brings rain. In turn, rainfall produces rivers and streams that move to the sea, providing power for water mills—early versions of today's water turbines—and today's hydroelectric generators.

Except for these few outright technological efforts to harness the sun's energy, little has been done through the years to put the sun to work. Until the Industrial Revolution, actually, little use was made of heat energy to provide working power for man. Until then he used his own muscles or the muscles of his animals—like horses and oxen—to produce work. The English language reflects that fact in the use of the word horsepower to measure the output power of machinery.

One exception is the harnessing of fire. But fire, except for military operations and for the fashioning of metallic products, was used only for heat and cooking until a few hundred years ago.

LET'S USE THE SUN!

Major energy transitions always bring about profound social change in the nations of the world. The substitution of coal for wood and wind ushered in the industrial era during the nineteenth century. Petroleum, in turn, revolutionized mankind's approach to travel, restructuring our cities and shrinking our world.

Domestic oil production peaked in 1970 and has declined by about 400,000 barrels a day each subsequent year. Domestic natural gas production appears to have peaked in 1974. In 1975, the U.S. Geological Survey radically reduced its estimates of undiscovered oil and gas. Offshore drilling now yields a high percentage of dry holes.

The United States imports more oil today than it did before the 1973 oil embargo by the Organization of Petroleum Exporting Countries. Forty percent of all our oil is imported and 40 percent of those imports comes from the Middle East. This dependency nurtures anxiety in high places. Pressure is mounting in the United States to shift more rapidly toward some new major source of energy.

Now, at the twilight of the oil age, the world faces another energy transition in the certain knowledge that it will reshape tomorrow's society.

We will always need fuel. Where can we get this precious energy? The answer is obvious.

From the sun.

We can become an entirely solar-powered society, the first in the history of mankind. Solar energy batteries, thermoelectric systems, and wind power can supplement direct solar heating and cooling systems.

The advantages of solar energy are many. It produces no radioactive wastes. Atomic bombs cannot be made from sunshine.

And what a change in our society a conversion to solar energy would make!

Desert areas would become vast fields of mirrors harvesting solar energy for generation into electric power to supply high-rise cities.

Suburbs would be spread even further outward from the cities in order to allow them the proper unobstructed prospective to collect the sun's energy to heat, cool, and run their homes.

Local zoning laws would have to be amended to protect an individual's "sunshine rights," to prevent a neighbor from building a higher structure that might cut out a man's source of power.

Tax laws would have to be revised to help the homeowner install solar energy heating and cooling systems—perhaps even solar cooking and refrigerating appliances.

The use of solar energy to run manufacturing plants *and* homes would free fossil fuel, and likewise release petrochemical feedstocks for the manufacture of materials like plastics, used now with increasing frequency in building and structural products.

Substituting a man-made plastics material for a natural product like lumber would bring about an environmental revolution. Plastics —manufactured with much less energy than steel, aluminum, and other structural metals—could possibly be made by solar energy.

SOLARON CORPORATION
Single installation of solar collectors serves two families in this duplex contemporary house in Fort Collins, Colorado. House may well be American dream of tomorrow.

Most of the pollution now caused by factories belching forth residue from combustible fuels would be nonexistent.

Solar batteries charged by exposure to the sun are technologically feasible right now. Quite possibly solar batteries could run our automobiles in the future, freeing mankind from smog and poisonous effluent, and clearing the air for the sun to shine more freely and provide even more potential energy.

Indeed, by returning to the sun for our source of energy, man can ride out the energy crisis and look forward to a cleaner, unpolluted earth and a brighter, more cheerful tomorrow.

2

Pioneers in Solar Heating

The cave was man's first natural living quarters. During the day the sun's direct heat provided enough warmth for him in his dwelling place. When he discovered the use of fire, he warmed his house with it at night when the sun had set.

Once out of the cave, man carried his fire with him into his newly built house. During the day the sun's rays warmed his dwelling, and at night he used the fire for comfort. By cutting holes in the house he let the sun shine in during the day, covering the holes with translucent material to let in only heat and light.

Except for that infinitesimal amount of solar help, man turned his back on the sun for many centuries. Yet ancient man, in spite of not using the sun's heat to warm his home, understood the sun and knew how to make use of its heat and light.

As early as the third century B.C., Archimedes—a philosopher, mathematician, and developer of ancient military hardware (he invented the catapult)—used curved glass lenses to concentrate the rays of the sun on Roman ships closing in on Greece during the siege of Syracuse. Mounting these enormous lenses on movable hinges so that his men could keep the rays of the sun directly focused on the sails of the approaching siege ships, he managed to set the sails on fire before the ships got within bow-and-arrow range of the shore.

MIRRORS TO CATCH THE SUN'S HEAT

For many centuries, there is no further record that solar energy was used by scientists—until 1747, when a Frenchman, Georges de Buffon, wrote a long scientific treatise called *Histoire Naturelle.* In it, he described how he had constructed a machine composed of almost 200 flat mirrors, turning them to concentrate the sun's rays on a piece of wood 200 feet away. He succeeded ultimately in setting the wood on fire.

Using the same contraption, he melted lead at 130 feet, and silver at 60 feet. This was the first time the sun's energy was actually used to achieve the melting of a metal.

SOLAR STEAM ENGINE

Back in 1698 in England, an enterprising inventor named Thomas Savery had patented what he called an atmospheric steam engine. Theoretically, it was a good machine, but it didn't work too well.

An improved version of the same engine was developed in 1712 by another engineer named Thomas Newcomen. This was a more practical version of the Savery machine, and it actually succeeded in pumping water out of a deep well.

The culmination of all this activity to harness hot steam was the invention of James Watt's steam engine in 1781, complete with a separate condensing chamber for recycling, an air pump to bring steam into the chamber, and with insulation used to cover parts of the engine to keep the heat in.

The way was now clear for man to use heat to drive machines, rather than the muscles of his animals or of his fellow man.

An early model of the heat engine, in fact, did make use of the sun's energy. Solar rays were focused on a simple boiler, the water was heated until it vaporized into steam, and the hot steam then ran the engine. But because the sun did not always shine, and because heat provided by burning wood and coal—which had been used increasingly during the preceding years—produced a more intense and quicker energy, the experiments with the sun's immense potential power were set aside.

It is interesting to note that up to this time about the only fuel man used to keep himself warm was wood. It was the only thing he knew that would burn easily.

Later on in the eighteenth century another scientist named Antoine Lavoisier designed a machine to "test the purity" of solar heat. The machine was made up of two large glass lenses mounted several feet apart. One lens was 52 inches in diameter, the other 8.

By directing the sun's rays through the large lens first and then the small one, Lavoisier was able to achieve a searingly hot temperature of 1,750° centigrade—3,182° Fahrenheit.

In another notable experiment, he succeeded in melting platinum. One of Lavoisier's chief claims to fame was his ingenious explanation of the theory of combustion, putting to rest the outmoded theory of phlogiston current at the time.

Lavoisier had better luck with inanimate objects than with his fellow man. As a tax collector during the time of the French Revolution, he was put to death on the guillotine by the peasantry.

Steam Power from the Sun The chief energy-producing fuels of the Industrial Revolution—which actually came about as a result of Watt's development of the steam engine—were coal at first, and later oil. Early locomotives were fueled by means of burning wood. So were many early boats.

But engineers never really discarded solar energy. The enticing thing about the sun was that its energy was free and easily captured.

In the 1860s another Frenchman, Augustin-Bernard Mouchot, decided to try his hand at a completely solar-powered steam engine, which he perfected soon and tried to finance. He showed the engine all around France, and later on tested it out in Algiers.

Mouchot developed a water-distillation plant which was completely successful and which was a very forward-looking desalination device similar to present models. Napoleon III's government, however, decided the machine could not be developed for a reasonable profit. When the government abandoned Mouchot, he gave up.

A decade later, in northern Chile, a solar distiller, or still, was designed to effect the same results as Mouchot's machine. It extracted salt from brine and brackish water to produce pure water. The still, constructed of glass panes in an early style of today's greenhouse, covered 51,000 square feet of area.

The sun's rays, penetrating through slanting panes of glass, heated troughs of water to high temperatures. When the water vaporized, it collected on the inside surface of the glass panes and condensed as it cooled. The condensation ran down the slanting glass into special channels as purified water. The working still provided up to 6,000 gallons of fresh water per day. Nothing came of it, however, and it was eventually abandoned.

In 1883, Swedish-born American inventor John Ericsson built a huge 11-by-16-foot solar engine. It was composed of a rectangular parabolic glass collector to catch the sun's rays. When the water inside was heated sufficiently to turn to steam, the steam drove a piston with a 6-inch bore and 8-inch stroke. The engine could deliver 1 horsepower for each 100 square feet of collector.

In Bombay, in 1876, a 2.5-horsepower steam engine was designed, heated by a hemisphere of 10-by-17-inch mirrors. The entire collector space measured 40 feet in length. The engine pumped water successfully.

About 1880, in France, Abel Pifre built a solar engine which was the first ever used for a commercial purpose. Fired by a parabolic collector 100 feet square, the steam engine produced ⅔ horsepower. It operated a printing press which printed a newspaper called *Le Journal Soleil.*

One of the first flat-plate collector solar engines was built by Charles Albert Tellier in France. The collecting area was 215 square feet. Instead of steam, air, or water, the driving fluid was ammonia.

In 1901, A. G. Eneas built a solar engine to pump water for an ostrich farm in Pasadena, California. His solar collector was built in the shape of a large truncated cone and contained 150 square feet of collector area. The engine produced 4.5 horsepower.

Sun Engines Intrigued by Tellier's flat-plate collector, two engineers named H. E. Willsie and John Boyle, Jr., built several sun engines in Illinois and Missouri. In principle, these sun engines trapped heat in water that flowed through glass-covered basins. The heat was then used to power water-ammonia engines.

By 1905, the pair were in Needles, California. There they constructed a 600-square-foot solar collector which trapped heat to operate a slide-valve engine running a water pump, a compressor, and two circulating pumps.

In 1908, they made a larger one of 1,000 square feet of collector area. Heat trapped in the collectors gave them a boiler pressure of 2.5 pounds, driving a sulfur-dioxide engine that produced up to 15 horsepower.

These engines were costly to construct—about $164 per horsepower, or four times the cost of a conventional steam plant. Even though the fuel was free, no one wanted to sink money into the newly risen solar-engine industry.

By 1907, another engineer named Frank Shuman came on the scene, determined to use flat-plate collectors for his engines. A 1,200-square-foot collector produced about 3.5 horsepower. The engineer then interested a British company in underwriting him and produced a pumping plant near Cairo, Egypt, that achieved 52.4 to 63 horsepower throughout its operation.

The plant was constructed of seven curved parabolic reflectors that concentrated the sun's heat on boiler tubes to produce steam at the focal point. The tubes were painted black. The reflectors, each 204 feet long, moved with the sun, following it automatically so that the intensity of the sun's power constantly hit the boiler tubes.

The steam that resulted ran an irrigation pump, but it could have driven a generator to produce electricity. The total collector area of Shuman's plant was more than 13,000 square feet.

Some years later an engineer named J. A. Harrington devised a solar machine which focused sunlight onto a boiler that produced steam to run a water pump. The water ran up into a 5,000-gallon tank, from which it then ran down through a water turbine to operate a dynamo that supplied electricity for the lights in a mine.

Dr. G. C. Abbott in 1936 designed a sun steam engine that delivered ½ horsepower. It was exhibited at the International Power Conference in Washington, where it furnished the power for a nationwide radio broadcast.

During that time, a French scientist named Georges Claude was conducting experiments aimed at tapping sea thermal energy. His idea was to run a steam turbine by combining the difference in water temperature at the surface and the bottom of the sea.

In 1931, Dr. Bruno Lange demonstrated what was called a photovoltaic solar-power battery. The idea of the solar battery was to trap the sun's energy in battery form to store it for future use. His battery consisted of a sandwich layered by copper oxide, silver selenide, and an unknown ingredient. When exposed to the sunlight, the battery could power a small electric motor.

In 1935, Professor Colin Fink of Columbia University devised a battery consisting of layers of copper and copper oxide. When exposed to the sun, it would run small motors.

Both Shuman and Harrington had constructed solar plants capable of generating electricity, but the output of those earlier versions were modest efforts compared with today's multimegawatt electric-generating power plants.

There are a number of solar power plants on the drawing boards today, many of them with enormous electric power potential. Some fairly large ones have already been constructed. There is an impressive sun furnace in France which can produce about 1,000 kilowatts.

In fact, scientists have already begun to apply solar concepts to power generation. There are also a number of other approaches to the conversion of the sun's energy to electric power ranging from photovoltaic and heliothermal "solar farms" to sea thermal-energy plants and huge, orbiting solar plants in space.

Those nearest fruition are land-based designs that will function best in the sunny Southwest.

HOME HEATING: THE PIONEERS

Even with all this activity in the solar-energy field, it was not until thirty years ago that basic research began on the development of methods to heat the modern home with solar energy. Money from a fund was given to the Massachusetts Institute of Technology and Harvard University to carry out experimental work on heating a house with solar energy, on the construction of flat-plate collectors, and on photochemical possibilities of heating.

In the 1940s practical developments were carried out by Dr. Maria Telkes and Dr. George O. G. Löf, both of whom did some of their work for M.I.T. Guggenheim Fund money, and support from the Rockefeller Foundation, starting in 1955; initiated a program of research at the University of Wisconsin on the use of solar energy.

Harold R. Hay began work in the 1950s in New Delhi, India, trying to devise a house that could be heated and cooled by solar energy. His innovations are discussed more fully in Chapter Nine.

The Hay Roof-Pool House Briefly, what Hay did was to devise his own type of solar collector—a simple pond of water on the flat roof of a house. By controlling the amount of insulation with a moving panel, he was able to control the warming and cooling of the water: the warming by the sun, the cooling by the blowing of air across its surface.

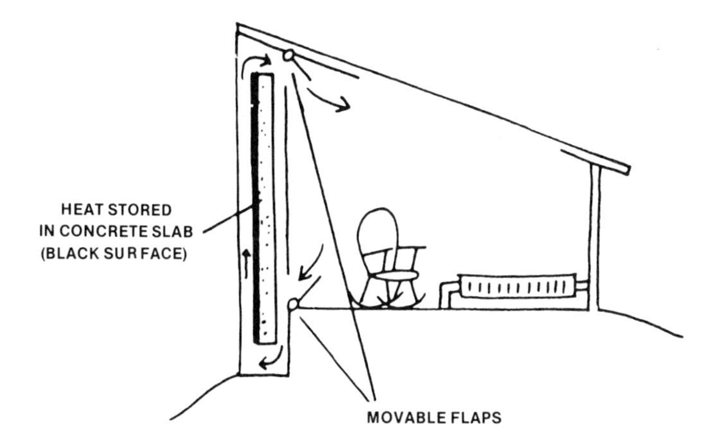

HUD

Drawing shows typical Trombe-type house, in which black-painted concrete wall insulated by glass cover traps air and sends it into house.

His original plan has been developed into his Skytherm patent. There are numbers of his installations in the country.

The Trombe Glass-Wall House In the meantime, Professor Felix Trombe was working on his own design of a solar-heated house in France. His was likewise a simple idea in concept. The house was constructed so that the south wall was a windowless and doorless surface built of thick masonry.

A glass partition, constructed 10 to 18 inches away from the outer surface of the wall, allowed the air in between to be warmed when the sun shone through the glass onto the black-painted wall.

As the wall warmed, the air trapped between it and the glass flowed upward into a storage zone inside the house. It remained there in storage at the top of the house, insulated and trapped, ready for use at night on demand.

The Thomason Solar-Collector House Along about this same time Harry E. Thomason built his first solar-heated and -cooled house in District Heights, Maryland, a small suburb of Washington.

He used a simple, direct approach and employed technological

HUD

Top drawing shows essentials of Thomason-type house, with collector on roof and storage tank in basement. Bottom drawing shows backup furnace usually needed with space-heating system.

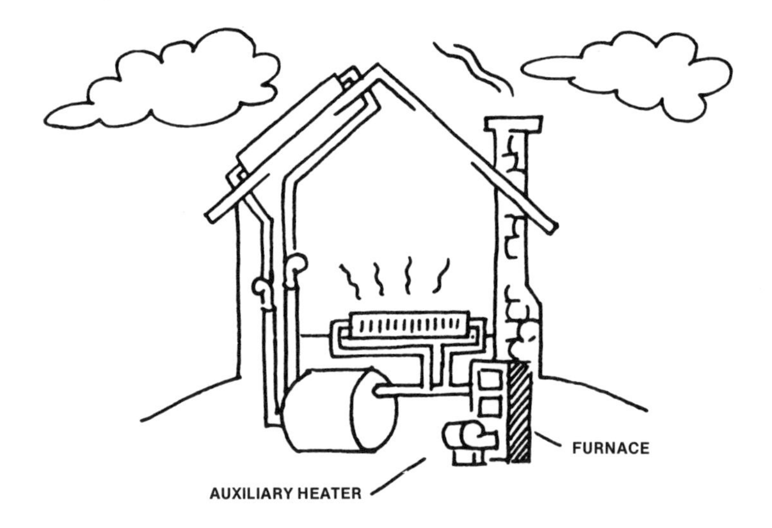

means that are used to this day by modern builders and designers in the collection and distribution of solar heat.

To catch the sun's rays, he built a solar collector out of corrugated aluminum, which he painted black. On the back of the aluminum he applied a flat piece of insulation. To the top he attached a large plate of glass and a large flat piece of plastic. He then mounted this collector on the south-facing slope of his pitched-roof house.

In the basement he used a 1,600-gallon water tank to store the warmed water. Connecting the roof collector and the storage tank with regular plumbing pipes, he ran one pipe to carry water to the top of the collector, and another to bring it down from the roof to the tank. In the top pipe he drilled holes to allow the water to trickle down over as much of the corrugated aluminum surface as possible.

Around the storage tank he packed several tons of crushed rock. The stones became heated as the storage tank was warmed by the collected water. Over the rock pile he placed a blower, which brought in cool air from the house, passed it over the hot rocks, and sent it back into the house warmed.

He installed a backup heater in case the sun was obscured by clouds or haze. The first year he let the solar heater run, he did not need the backup until well into December—and then it ran for only a half hour or so.

SOLARON CORPORATION

Air-medium solar collectors installed on mansard-type roof and south-facing wall give winter heating to this Colorado mountain home.

Later, he decided to reverse the system and make himself an air cooler out of the same arrangement. On the north slope of his roof, he mounted an aluminum collector, and attached it to the storage tank. The collector ran water across its surface only at night, was cooled by the night air, and then was returned to the storage tank. In the morning the house began to heat up. The blower turned on and sucked the hot air out of the house, blew it across the cooled rocks, and sent it back up into the house several degrees cooler.

The arrangement didn't seem to work so well. Later on, and in another house, he built a new solar cooler. Here he cooled the rocks by means of an electric compressor. The compressor not only blew the rocks cool but also sucked out a great deal of the humidity that came in with the wet warm air.

This system worked much better for the cooling of the house's interior, because it removed some of the humidity from the air as well as the high temperature.

Thomason's method is patented—he worked in the U.S. Patent office at the time he developed it—and he sells plans for the system, which he calls Solaris, even today.

In spite of these isolated instances of successful solar houses, and in spite of the fact that in postwar America all the technology for the complete solar house, run totally by energy captured from the sun's rays, was ready to be put together—in spite of that, nothing happened.

Nobody wanted solar heat.

Frankly, it was easier and cheaper to heat houses with conventional oil-fired furnaces. If oil was too expensive, then gas-fired furnaces could be operated much more cheaply. And, for the affluent person, or for one who lived in an area where cheap electric power was available, the electric-resistance heater was an excellent, dependable servant.

But there were those who still persisted in touting solar heating and cooling for the home. Yet it wasn't until inflation and the energy crunch, exacerbated by the price hike instituted by the Organization of Petroleum Exporting Countries, plus the shortage of natural gas aggravated by the disastrous 1976–77 winter that the need for solar energy for the home really arrived.

Now, once again, man looks out the window of his house and thinks about harnessing the sun's power. The technology is all there, borrowed from hundreds of years of scientific discovery and experimentation. All it takes is the desire to do something about it.

Today, finally, someone *is* doing something about it.

3

How Solar Collectors Work

The conventional home heating system includes a furnace where fuel is burned, a boiler or tank where heat is stored for immediate or future use, and a system of pipes or ducts to carry the heat by some medium to where it is needed.

A solar heating system varies from this only in that no fuel is needed. In place of fuel, the solar system utilizes the heat of the sun.

The conventional solar heating system includes a collector which gathers in the sun's heat, a storage zone where heat is stored for immediate or future use, and a hookup of pipes and ducts to distribute heat where it is needed.

The parts of the solar system are divided for convenience's sake into two subsystems—the collection subsystem and the delivery subsystem.

THE SOLAR COLLECTOR

The key to any solar heating or cooling system is the solar collector. Taken in its simplest and most primitive form, an ordinary closed room with a closed glass window facing the south on a hot bright day is a direct collector of solar energy. The sun's radiation enters through the glass and heats the objects in the room. The air is in turn warmed by conduction from the objects.

A simple solar collector is similar to a closed room with a southern exposure. Because the collector has to be in position for maximum absorption of the sun's energy, it is usually placed on the roof of a house or on the south wall, where the sun's rays fall at a maximum angle for warming purposes.

When any object is exposed to solar radiation, its temperature rises until it equals the temperature of the objects around it. Surrounding colder air will cause heat to radiate from the object. So will heat move from it to materials touching it by conduction.

The temperature gains depend on the intensity of solar radiation and on the absorption ability of the object in question. It also depends on the amount of heat radiated out from the object after absorption.

If the object being warmed is contained within an air trap of some kind, the radiation from the object warms the air in turn. If the air trap is formed by a glass case, for example, the object and the air will continue to become warmer inside the glass trap.

As an object increases in temperature, its radiation potential increases. At around 212° F. and higher, heat with the smallest wavelengths—from 8 to 10 micrometers in the infrared area—is at a maximum. Energy at that wavelength passes through transparent material like glass, but if the glass is doubled, the transmission is decreased greatly. Sunlight with wavelengths of less than 2.5 micrometers passes through any number of layers.

It is because of this property of light that most solar collectors are built with the same design and components.

TYPES OF SOLAR COLLECTORS

There are actually two different kinds of solar collectors—one that is composed of a receiving surface covered with sheets of glass or plastic, and one that is composed of mirrors or lenses which focus solar radiation onto a smaller area for more intense heat.

Solar collectors used in residential dwellings are mostly classed as flat-plate collectors which do not focus or bend light but absorb it directly from the sun.

In turn, flat-plate collectors come in two models.

The *air-medium collector* absorbs the sun's energy and captures its heat in air trapped in the collector box.

The *fluid-medium collector* absorbs the sun's energy and captures its heat in a fluid that circulates through the box.

COMPONENTS OF SOLAR COLLECTORS

The residential solar collector is usually built in the form of a large flat rectangle, although some are formed in cylindrical shapes and other configurations. The flat-plate collector is usually about three feet wide by six feet long, and less than a foot thick. Its shape makes it easier to mount on a roof or wall.

It has three main components: the cover; the absorber plate; and the insulator back.

The cover comprises two panes of glass or clear rigid plastic through which the sun's radiation penetrates into the rectangular collector box, heating the air or fluid inside.

The absorber plate forms the bottom of the box to prevent the heat from the solar collector from seeping through the roof or wall to which it is secured.

In the air-medium collector, the heated air can be trapped in the space between the double panes of glass or plastic and the absorber plate, or underneath the absorber plate. In either case, the heated air is ducted out to the storage system.

PHELPS DODGE BRASS COMPANY

Typical solar collector has glass covering, black-coated absorber plate of aluminum beneath, and built-in tubing to carry fluid medium through plate.

KALWALL CORPORATION

Not all solar collector covers are transparent. Translucent fiberglass lets in varying amounts of light, as in this collector. Absorber plate and tubing for fluid medium stands beside cover. Picture below shows collectors mounted on roof of residence.

In the fluid-medium collector, the absorber plate is attached directly to a separate component—the conduit.

The conduit is the heart of the fluid-medium flat-plate collector. It is composed of metal tubing and fins of various designs. The conduit carries the fluid medium through or under the absorber plate, where it collects the heat and conveys it to the storage zone.

The Air-Medium Collector The design of the air-medium solar collector, which delivers heated air to the storage zone, is actually a simple imitation of the Trombe-type house described in Chapter Two.

Diagram shows details of typical fluid-medium solar collector absorber plate. Fluid flows through S-shaped rectangular copper tubes.

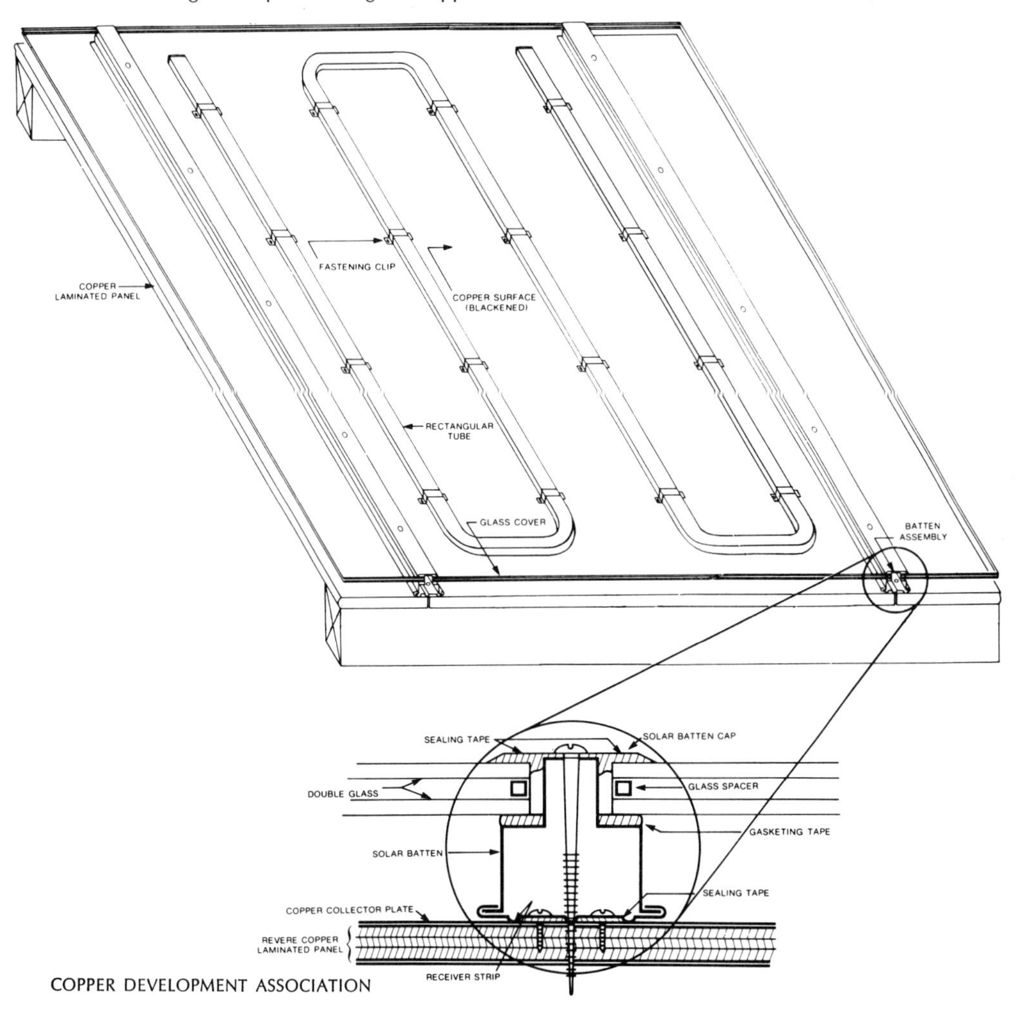

COPPER DEVELOPMENT ASSOCIATION

TYPES OF CIRCUITING

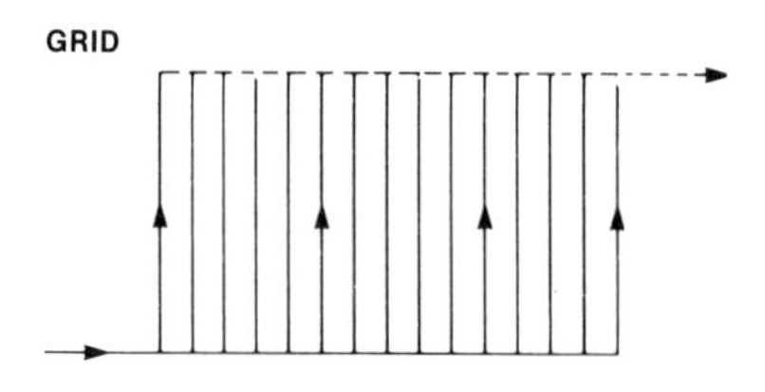

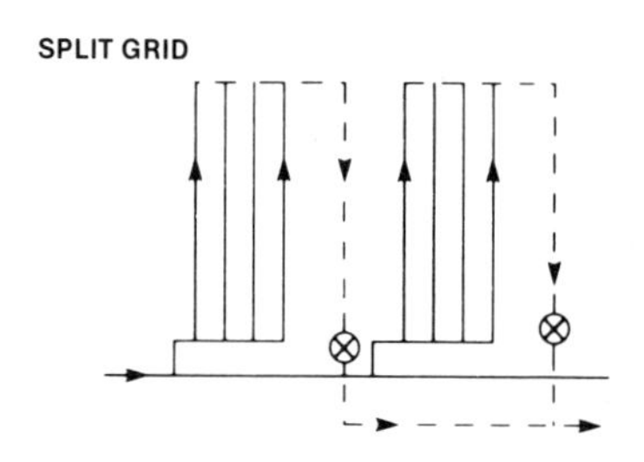

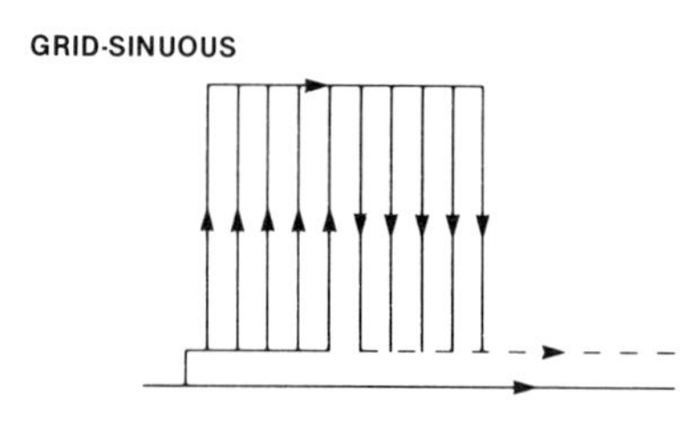

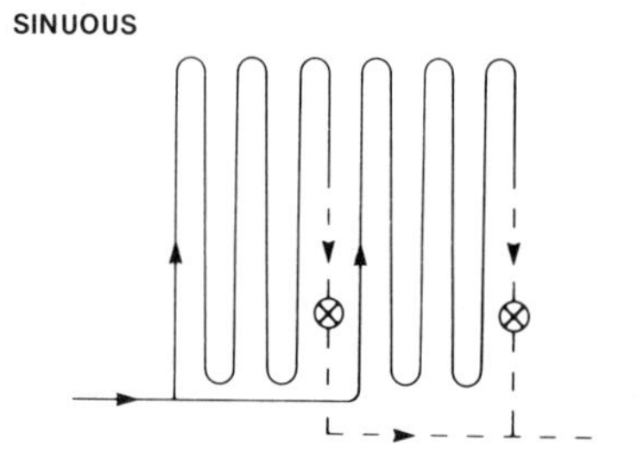

Diagrams show four different circuiting configurations for conduit carrying fluid medium through flat-plate collector. Conduit may contain water or water mixed with antifreeze.

REVERE COPPER AND BRASS INCORPORATED

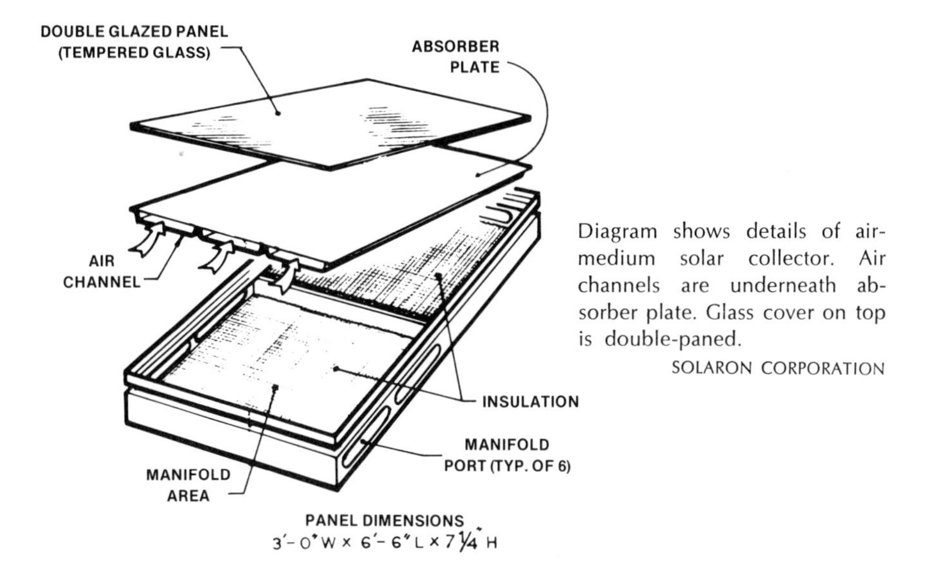

Diagram shows details of air-medium solar collector. Air channels are underneath absorber plate. Glass cover on top is double-paned.

SOLARON CORPORATION

Like the Trombe house, the air-medium collector is composed of a thick slab of insulation on the bottom, an absorber plate painted black resting on that, and double panes of glass or plastic on the top acting as cover.

Air from the interior of the house circulates through the collector between the black-painted surface of the absorber and the enclosing cover. As the air is heated by the sun, it rises, and is drawn in through ducts at the top to a storage or distribution zone inside the house.

The Fluid-Medium Collector The fluid-medium collector is based loosely on the "roof pool collector" described in Chapter Two. From bottom up, it is composed of a thick slab of insulation to keep the heat inside the collector box. Directly above the insulation backing is the absorber plate. Imbedded in the absorber plate is a circuit of pipes, running either vertically up and down the collector box, or in S shapes and other configurations. The circuit carries the liquid through the absorber plate for maximum heat transfer.

As the sun shines through the cover, the black absorber plate inside the box warms up. The fluid passing through the conduit in the absorber plate collects the heat and carries it through pipes to the storage zone.

THE FLAT-PLATE COLLECTOR COVER

The transparent cover of the solar collector can be made either of glass or plastic, depending on the desire or needs of the manufacturer or buyer. Durability, translucence, price, and various other factors can be a consideration.

The Glass Collector Cover As a material, glass holds up well through the years. However, it is fragile and will shatter if struck by rocks, birds, or other moving objects. In areas of high-velocity winds—hurricanes, gales, and other violent movements of air—even shatterproof glass may break.

Since a flat-plate collector must be completely unsheltered to catch the sun's rays, there is no way of protecting it against moving missiles.

Yet in areas where there is snowfall, glass at the correct angle of declination on a pitched roof will successfully melt average snowfalls and may be able to sustain more weight than plastic materials.

KALWALL CORPORATION

Double-paned translucent plastic sheets of off-beat vertical collector does let in sunlight during day. At night thousands of styrofoam beads are blown into space to form tight insulation wall.

The Plastic Collector Cover Some manufacturers of flat-plate collectors prefer plastic to glass. The reason for the use of plastic in some cases is that it can be formulated into many different kinds of materials, each with specific properties.

The chemical engineer can build in specific properties by creating the plastic formulation to his own desires. The versatility of plastic is inherent in its composition—it is a complexity of simple carbon-based molecules linked together by chemical technology.

Durability, transparency, insulative ability, corrosion-resistance, tensile strength—almost any value desired—is simply built into the formula. Many such formulations have been devised already in an attempt to produce the optimum solar collector cover. None has yet been found combining properties containing only advantages and no disadvantages.

The advantage of plastic in covers is its inexpensive cost. Certain formulations deteriorate in sunlight and in heavy weathering conditions. But new combinations have been developed to resist weathering and sunlight.

Other properties of plastic that are important are its ability to take on a water film, the ease of removing dust, the influence of temperature on deterioration, and the loss of transparency for sunlight. Some plastic films can be heat-sealed and others cannot.

Transmission of light through plastic depends on the refractive index of the material and the reflections at the two air-plastic interfaces. The absorption within the material is less than within glass because the films of plastic are much thinner than the glass plates.

Sunlight consists of both direct radiation and scattered radiation. The transmission may depend in part on the ratio of the two.

One new combination of glass or acrylic plastic coated with Teflon is being tested out on solar panels. A collection of 8 to 30 percent more heat has been recorded by means of the coated panels. It is the low refractive index of the Teflon film that allows more light to pass through and in turn affords more heat in the collector box, according to the developers at the Du Pont Company.

The ability of the coating to let through more light enables the solar collector to pick up heat earlier in the morning and later in the afternoon than ordinary panels can.

On hazy days, the coating picks up more of the scattered light available in the atmosphere. Actually, the coating makes its greatest gains on cloudy days. Such a panel reduces the need for backup heating in the solar system during overcast periods.

This plastic coating is only one of many experiments under way. A team at the Solar Research Laboratory, Massachusetts Institute of Technology, is working on a type of coating that will make glass an even better collector of solar energy than it is now.

Scientist Day Chahroudi and his associates, John Brooks, a chemist, and Sean Wellesley-Miller, an MIT professor of architecture, are trying to devise a coating that will make glass transmit more radiation directly and reflect less of it into the atmosphere.

The problem with any collector is that, as the sun continues to shine and heat the plate, the collector box becomes too hot in the middle of the day, and a great deal of collectable energy radiates back into the atmosphere and is lost. At night the heat that has been absorbed during the day and not transmitted to the storage area is also lost by radiation back into the atmosphere.

Chahroudi and his associates have come up with what they call a "solar membrane," which they claim retains about 90 percent of the heat needed to operate a solar house. The average total residence heating done by most solar units is now about 75 percent.

In a closely allied experiment, the group is at work on a material that will prevent the solar collector from overheating in the day when the sun's heat is at its highest, causing waste by reradiation. They have come up with what they call "cloud gel," a membrane that will change from clear to opaque white at a preset temperature —say, of 80 degrees. It is, unfortunately, very expensive at the present time to produce this special material.

THE ABSORBER PLATE

As for the black-painted metal absorber plate, it is usually a flat piece of metal which absorbs the sun's light, transforms it to heat by absorption, and in turn radiates heat into the air surrounding it between the cover and itself.

The absorber plate in a fluid-medium collector is usually constructed differently because it must contain the conduit, or piping circuit, that carries the fluid medium to and from the collector.

There is no end to the variety of design in the absorber plate, depending on the wishes of its designer and manufacturer.

Copper and aluminum are the two most commonly used types of metal. The shapes and configurations of the absorber plate are endless.

ALUMINUM ASSOCIATION

This fluid-medium solar collector features S-shaped extruded aluminum tube serpentined into a flat-plate absorber painted black.

The type of absorber material used in the collector box is chosen for the particular advantage and/or disadvantage the manufacturer conceives after considering cost, durability, resistance to corrosion and ultraviolet rays, heat absorbency, heat retention, weight, and ease of construction.

CONSTRUCTION OF THE FLAT-PLATE COLLECTOR

Weight and strength as well as appearance are definite considerations in the design of any solar collector. The box must be assembled in as rigid a form as possible, so that it won't affect the structural shape of the house's roof, of which it is actually a part.

The average collector must be airtight, of course, with the only ingress and egress connected to pipes or ducts carrying the fluid or air to and from the collector box.

For esthetic reasons, the gleaming black collector plate presents problems to architects trying to design new houses heated by solar energy. The placement of these plates must usually be on the south side of the structure, and that site should be free of any blockage by hills, trees, or other structure or shapes.

Most manufacturers produce surface-mounted modules that serve for both fluid-medium heating and air-medium heating collectors. There are many different designs with considerable variation in appearance available to the architect.

MOUNTING THE FLAT-PLATE COLLECTOR

The siting of the flat-plate collector is extremely important, since it must catch the maximum amount of sunlight possible. The angle at which the collector is tilted to the sun depends not only on the use to which the collector is to be put but to the geographical location of the building site.

For the collection of energy for space heating only, the optimum angle of inclination from the horizontal can be calculated by adding 15° to the local latitude. In other words, for a heating system installed in a home at about 42° north latitude—a line running through Cedar Rapids, Iowa, for example—the collector should be tilted at an inclination of 57°.

For a year-round solar heating and cooling system that includes hot water heating as well, the best and most efficient angle is exactly the same as the local latitude.

In fact, the angle can vary by plus or minus 10 percent of the optimum inclination with little change in the efficiency of the collector.

WHERE TO PUT THE COLLECTOR

The possible blocking of a solar collector by a tree or a structure will cause it to perform less efficiently than it should. The solar collector must have an unobstructed view of the sun for at least six hours of the day. At least half of this time should be on either side of solar noon.

Careful checking should be done to determine whether or not solar collectors will be shaded from the low December sun by nearby terrain, buildings, evergreen woods, other rows of solar collectors, chimneys, or whatever.

It is even possible to mount the solar collector on a vertical south-facing wall if it is to be used for space heating only. However, a greater collection area is needed with vertical installation. Some

Workers assemble solar collector plates above. Black-painted plate is mounted on top of conventional sloped roof at left.

ALUMINUM ASSOCIATION

of the sun's rays are deflected and lost because of the angle of penetration.

The truth is, vertical installations are not very effective for domestic water heating, since the amount of solar energy striking the collector during the summer when the sun is overhead is minimal.

The best orientation for the flat-plate solar collector is slightly west of true south. The sun's rays produce more heat in the collectors after high noon. Solar collectors can be placed as much as 20 degrees from due south in either direction, and still function well.

If it is not possible to orient the collector box in the due-south position, it is wise to increase the amount of collection area. Incidentally, "due south" is *true* south, not magnetic south. Variations in the compass must be calculated accurately in order not to lose efficiency.

HOW HOT DO FLAT-PLATE COLLECTORS HEAT?

The temperature attained by the average collector depends on several factors, among them the amount of the sun's rays striking the collector and the amount of heat transmitted to the storage zone.

On a clear day it is possible for the average flat-plate collector to heat the medium to between 100° and 150° F. above the outside temperature. That means that the solar collector should be able to produce water measuring over 200° in the storage zone.

On a hot summer day, temperatures above the boiling point are often produced. During periods when there is no flow of fluid or air through the collector, it may attain a temperature of up to 400° F.

HOW DURABLE IS A FLAT-PLATE COLLECTOR?

The strength and durability of the solar collector must be considered carefully when comparing costs and types. In sections of the country where there are severe hailstorms, or in places where there is a likelihood of vandalism, it is sometimes wise to protect each collector by covering it with half-inch hardware cloth. The addition of this transparent cover will reduce the collector area by about 10 percent.

The ordinary solar collector usually comes guaranteed for five years or more. With such a guarantee, the collector should last for about twenty-five years, since there are no moving parts to worry about.

The glass or plastic cover of any solar collector will get dirty. The fall of rain will naturally wash off the surface to a degree. The owner may have to wash the cover once a year or even more in order to retain optimum performance.

In snowfall regions, the solar collector should be inclined at a sharp angle—50° or more—to keep snow from accumulating. However, if it does stick, it will not cut down performance too much. Snow is transparent enough to let in sunlight if it is less than 6 inches deep.

With the warming of the collector, the heat inside will speed up the melting of the snow layer on the glass, letting it slide off. If a thick blanket of snow falls, the homeowner need only circulate hot water through the collectors from the storage tank to melt it off.

In regions where freezing can be expected overnight, the collector system can be designed to allow the heat medium, if fluid, to drain into a container where it won't freeze. In some collectors, the fluid solution is a combination of water and antifreeze, which will resist freezing. In a collector in which water only is used, the owner can add antifreeze during a cold period to keep the solution from icing up.

THE STORAGE SUBSYSTEM

An essential part of the collector subsystem is the storage zone. Once solar heat has been collected by the roof panel, it is then circulated into the structure by means of pipes or ducts to the storage zone, where it is stored for immediate or future use.

The circulation subsystem that moves the heat from the collector to the storage zone is made up of pipes or ducts, carrying fluid or air. The medium is moved from the roof to the storage zone either by a pump or by gravity.

In addition to the pump, the circulation subsystem includes a differential thermostat, called a sensor. The sensor determines the proper time to move heat from the collector to the storage area.

For example, when the solar collector subsystem begins to produce less heat for storage in the late afternoon, or on a day when

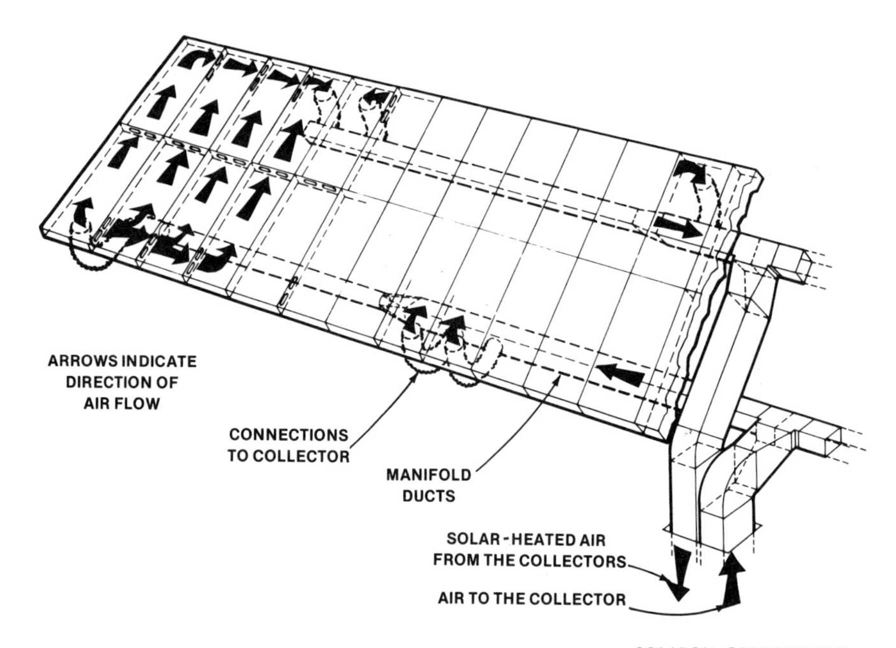

SOLARON CORPORATION

Drawing shows typical installation of air-medium flat-plate collectors, with ductwork leading into and out of the collector area.

the sun is obstructed by clouds, a sensor will turn off the pump that moves the medium through the collector.

It is a law of thermodynamics that heat moves from a heated area to an unheated area. Thus, if the storage subsystem becomes warmer than the solar collector, the air or fluid medium will not add to the heat in the storage area, but will actually take heat away from it.

The reverse is true in the morning hours. When the collector subsystem, which is not pumping, warms up enough to be several degrees hotter than the storage subsystem, the sensor activates the pump to circulate the collector medium.

Where the Heat Is Stored The design and construction of the storage subsystem of a solar heating unit are of vital importance to the efficiency of the system.

If the storage unit isn't properly designed and built, heat can escape and reduce the efficiency of the entire system. Because of the necessity of utilizing an enormous amount of space for storing heat, the problem of design and structure becomes a difficult one, especially in a small area like that involved in a residential unit.

Storage of Fluid-Medium Heat In a fluid-medium system, the storage subsystem is a hot water tank. Because of the amount of heat involved, the average size of this storage tank is about 1,000 gallons or more.

The storage tank is connected to both the domestic hot water system and the space-heating system. The fluid medium used in the collector is usually contained within a closed system that does not mix in any way with the water in the actual storage tank.

The typical storage tank is actually a heat exchanger. The typical heat exchanger is composed of two parts: a large tank in which water is stored; and a radiatorlike unit or coil of tubing inside.

The liquid medium from the solar collector is pumped down to the storage tank. It flows through the coil of tubing inside the tank, and, as it does so, it imparts its heat to the water there.

Because antifreeze has a much lower boiling point than regular water, it will usually be vaporized by the time it gets to the heat exchanger.

The fluid medium then condenses in the heat exchanger, imparting a great deal of heat to the water inside the storage tank. It then returns to the solar collector, cooled and ready to collect more energy from the sun.

Meanwhile the heat from the solar collector has been moved to the storage zone to warm the already heated water.

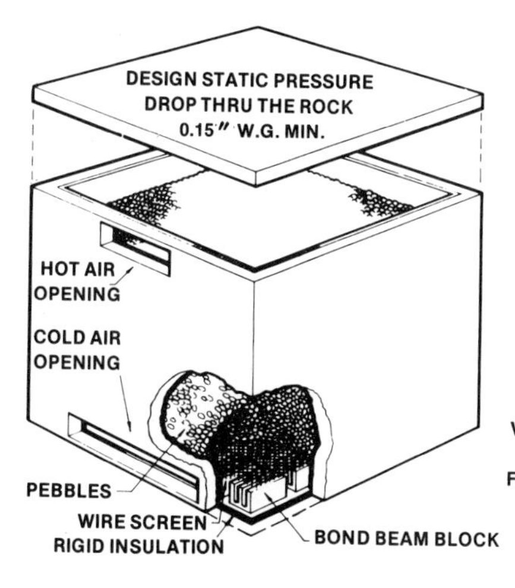

Drawing shows storage box for air-medium solar collector. Box contains pebbles about 3/4-inch to 1 1/2-inches in size. It can be made of plywood, reinforced concrete, concrete block, or other masonry.

SOLARON CORPORATION

Fluid heats quickly and cools quickly. It is not really a terribly effective storage medium at all, unless it is rigidly controlled by highly efficient insulation. For that reason, air-medium collectors have become popular for space heating in certain areas of the country.

Storage of Air-Medium Heat In the evening after the sun goes down, most of the earth cools off gradually. After several hours, the earth is cold, and so are plants growing in it. Rock, however, cools more slowly than earth or water.

Crushed rock has a very slow rate of absorbing heat and of radiating it away. Because of its physical ability to hold heat, it is used as a good and efficient method for retaining heat from air.

In air-medium solar systems, the crushed-rock storage zone can be kept in a large pit in the cellar, or even outside the house in an insulated hole.

The air warmed in the solar collector is ducted down into the pit where the crushed rock is stored. The warm air heats up the rock. It is pumped up again through the ductwork to gather more heat in the collector.

Space Needed for Storage Areas The basic problem with the storage of solar heat—both fluid-medium and air-medium—is the enormous amount of space needed for the storage water tank or the rock pile.

The average rock pile for a small house contains about 72 tons of rock—enough to present a problem for any architect trying to design a house around it.

As for the storage tank of water, the minimum size is a 1,000-gallon tank. In areas where there are fewer sunshiny days and more cold in the winter, the size of the storage tank can be up to 2,000 or more gallons of water.

4

Distributing Solar Energy

The second part of the conventional solar heating system is called the delivery subsystem. Its purpose is the final distribution of stored heat to the various rooms of the house where and as it is needed.

The delivery system involves sensors or thermostats which call for heating or cooling, pumps or fans which move the air or fluid to the areas calling for it, the recovery part of the system, where the heat is stored, and the backup unit for heating or cooling the house when solar heat does not provide enough energy during cloudy or overcast days.

PASSIVE AND ACTIVE SOLAR SYSTEMS

For convenience's sake, solar systems are divided into two types, depending on the amount of service they supply to the home and depending on the amount of nonsolar energy they use up in providing it.

Although a solar heating and cooling system which is run strictly by the sun's heat and does not require any other fuel or power source is theoretically possible, such a system is not yet in widespread practical use. Called a "passive" system, it does not use any other "active" type of power unit.

For practical purposes, the typical solar heating and cooling system now in use is an active type, with a backup unit of conventional fossil-fuel heat or electric-resistance heat to take care of the hours when the solar collectors and the storage units have run out of saved-up heat.

THE TYPICAL DELIVERY SYSTEM

The most complicated part of a solar heating plant is the control system needed to coordinate it not only with the sensors, telling it when and where to send heat, but to coordinate it with whatever type of backup system is used.

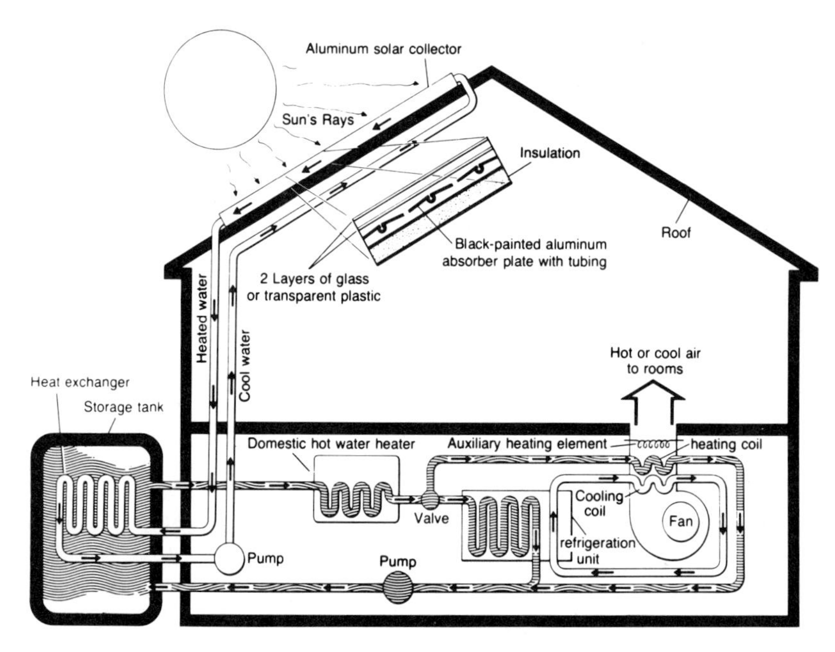

ALUMINUM ASSOCIATION

Typical solar system works this way: fluid medium from collector flows into exchanger, heating storage tank for domestic hot water and space heating.

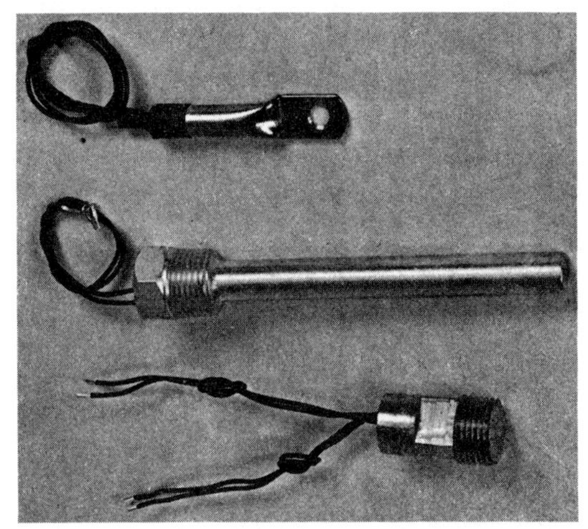

Sensors, or differential thermostats, are provided along pipe and ductwork of solar collector system to activate pumps and blowers when temperatures change.

RHO SIGMA INCORPORATED

THE RECOVERY SYSTEM

An essential part of the delivery system is the unit that accomplishes recovery of heat from the storage zone.

In an *air-medium* solar system, the heat is stored in a crushed-rock pit. To recover this heat, filtered air is usually blown through the crushed rock, where it gathers heat, and then is sent through ductwork for distribution throughout the house.

The ductwork built for a forced-air system, incidentally, is exactly like the ductwork for a conventional oil- or coal-burning forced-air heating system. In adapting such a system to solar heat, almost no changes need be made.

In a *fluid-medium* solar system, the heat is stored in a water tank. To recover this heat, the hot water is piped to a coil in an open air shaft. A fan or blower moves filtered air across the hot coil, where the air is warmed. This air is then introduced into ductwork and is distributed to the rooms where it is needed.

If the house is not provided with forced-air ductwork but is heated by a hot water or steam system, a slightly different type of distribution setup is necessary. The system is exactly like a conventional hot water or steam system.

In a hot water system, the warmed water is simply introduced into the hot water pipes and pumped into radiators, where the heat warms the fins or surfaces of the radiators and transfers into the air by conduction.

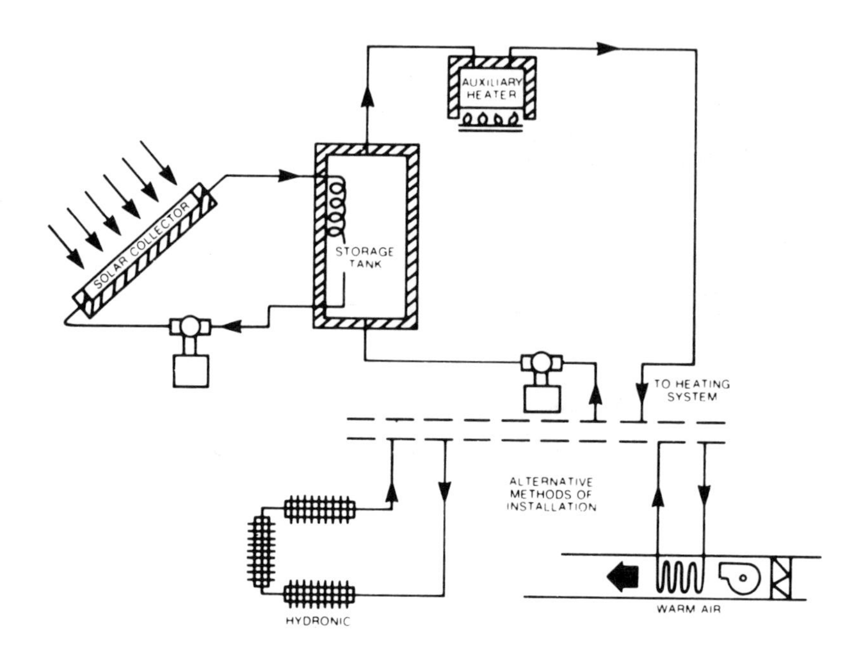

REVERE COPPER AND BRASS INCORPORATED

Fluid medium flows to storage tank, imparts heat through heat exchanger, and returns to collector. Backup heater takes over when system runs low on solar heat.

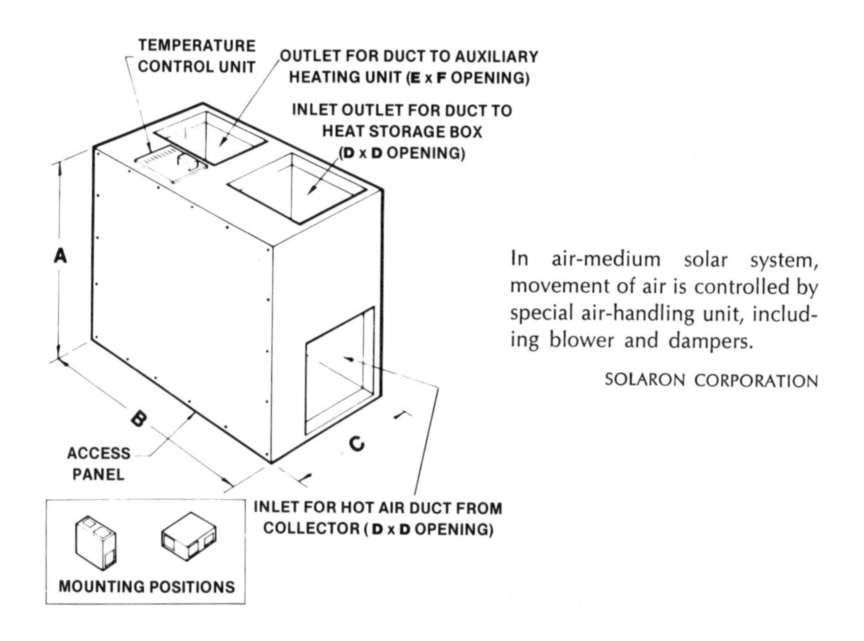

In air-medium solar system, movement of air is controlled by special air-handling unit, including blower and dampers.

SOLARON CORPORATION

The limitations of the hot water and steam system for distribution is obvious. Such a system cannot be used for conventional air-conditioning or cooling during hot weather.

BACKUP UNITS

By and large, most solar systems do need backup heating and cooling units for periods when solar energy is lost through cloudy days or when there are cold snaps demanding more heat.

The solar heating/cooling system can use an electric-resistance type of heater or a gas-fired type of heater for backup.

In general, almost any solar heating/cooling system must have electricity to activate the pumps and sensors that operate the delivery system.

These devices must be on call at all times in order to activate the backup heating unit when the storage subsystem is not warm enough to fulfill the demands of the house's thermostats.

What Kind of Backup Unit? In some southern states private residences use solar systems without backup of any kind. In warm climates, a solar home can get along with the occasional use of a fireplace on a crisp winter evening.

But generally speaking, most solar homes now in use need backups. Geographical location affects the type of backup system needed.

Houses north of 35° latitude—an east–west line which passes close to Little Rock, Arkansas—need backup systems for heating.

Below that line, houses need backup systems for air-conditioning or cooling.

Nevertheless, solar-assisted heating or air-conditioning systems are practical in every one of the fifty states, including Hawaii and Alaska.

Actually, even though it is generally cooler in high altitudes, the thin, clear air helps solar collectors retain more of the sun's energy. Called insolation, the amount of solar radiation received in Colorado, for example, makes it a very popular site for solar buildings.

Cloud cover is the most important factor to be considered when dealing with solar energy. The cloudier the area, the bigger and more expensive the collector installation will have to be.

Chapter Fourteen has a much more detailed breakdown of insolation, area by area and state by state.

ENTER THE HEAT PUMP

The recovery systems already discussed are based on developments made in coordination with extant systems of heating and cooling: forced-air furnaces; air-conditioning units; and hot water and steam systems.

Instead of utilizing the complex arrangements needed to coordinate a solar collection and storage system with an existing heating and cooling system, engineers are now working on what should prove to be the most efficient way to store and distribute solar energy: by means of the heat pump.

There is nothing new about the heat pump. As far back as the 1930s the heat pump had reached a degree of perfection. But because conventional heating and cooling methods were more in demand, the heat pump languished in development.

But today it is staging a comeback. An electric-powered device, it is used in almost a million homes to provide space heating alone. In fact, latest figures show that one out of every four new homebuyers in America will have a heat pump in his house for space heating.

It has made such rapid strides recently because it is an unusually efficient and effective machine.

What Is the Heat Pump? In the simplest sense, the device is a machine that moves heat not where it would usually go—toward a colder area—but in the opposite direction: from a cold place to a warm place. In other words, the heat pump makes cool areas cooler and warm areas warmer. (Heat allowed to move in its natural direction would make all places the same temperature.)

In summer the heat pump sucks heat out of the hot inside air and exhausts it outside. In winter it pulls heat out of even below-freezing outside air—there is always *some* heat in the air down to absolute zero, which is —460° F.—and uses it to heat the house's interior.

The heat pump is a close cousin to the refrigerator. In fact, its technology is strongly suggestive of that of the refrigerator. Most models use a Freon-type coolant, or antifreeze, to effect the movement of heat.

Compressor unit of typical heat pump is located outside wall of house. Pump provides heat in the winter and cooling in the summer by moving heat to where it is wanted.

WESTINGHOUSE ELECTRIC CORPORATION

Even though the heat pump runs by electricity, it is extremely efficient, and can deliver two or three times more heat to a home than is required to run it.

In a space heating installation, the heat pump can extract sun-derived heat from outside air and deliver it inside, acting as a kind of solar distribution pump.

How Does a Refrigerator Work? To understand how the heat pump operates, it is necessary to review the functions of the conventional home refrigerator. The "magic" of the refrigerator is based on application of several important thermodynamic principles of liquids and gases.

Every liquid, like water, has three physical states: the liquid state; the freezing state; and the boiling (gaseous) state. What we normally think of as gas—like natural gas—can be transformed not only into a liquid state, but into a frozen state as well. Water turns to vapor in its boiling state, to ice in its freezing state.

Different liquids have different freezing and boiling points. Water's boiling point is 212° F. (100° C.) and its freezing point is 32° F. (O° C.). Some liquids have different boiling and freezing points. R-12, which is the typical refrigerant used in commercial refrigerators, has a boiling point of −21° F. At ordinary temperatures, it is actually a gas.

When water boils at 212° F. the temperature of the water is reduced slightly. The reason for this is that heat must flow from the liquid to make up for the amount of heat taken away in the part that evaporates. This physical change is called the "flash" process. The practical effect of the flash process is the reduction of heat. Vaporization, then, causes heat to be taken away from the surroundings and makes them cooler.

When vapor is condensed back to water, the reverse process takes place. That is, heat is not taken away but is given off as part of the process of condensation. The practical effect of condensation is to give off heat and make the surroundings warmer.

One more fact of thermodynamics: An increase or decrease in atmospheric pressure changes the boiling point and freezing point of water. When atmospheric pressure is increased, the boiling point is raised. That is, the liquid boils at a higher temperature. When atmospheric pressure is reduced, the boiling point is lowered. That is, the liquid boils at a lower temperature.

And, when atmospheric pressure is increased, the *temperature* of a liquid is increased; likewise, when it's lowered, the temperature is lowered.

The conventional refrigerator works by combining these principles as follows:

It utilizes a gas like fluorinated hydrocarbons (Freon) with a low boiling point, alternately vaporizing it to a gas under reduced air pressure and condensing it to a liquid under increased air pressure.

The process of vaporization (boiling) extracts heat from the surroundings and cools the food and contents of the refrigeration area. The vapor is then compressed to liquid by raising the air pressure in the compression chamber. The heat produced during liquification is exhausted outside the unit and dissipates in the outside air.

The vaporization in the next step leads to a repetition of the cycle. The two operations—vaporization and condensation—become continuous. The heat inside the refrigeration unit, in effect, is exhausted from it during compression to produce cooling inside.

How the Heat Pump Works This cycle of vaporization and condensation is similar to the cycle employed by the heat pump to move heat. Using the same processes as a refrigerator unit, the heat pump alternately vaporizes and condenses a refrigerant, causing circulating air to be warmed and cooled. In winter the warmed air is circulated to heat the house. In summer the cooled air is circulated to cool the house.

In conventional installations, the heat pump has five basic components: an indoor coil; and outdoor coil; a reversing valve; a compressor; and a refrigerant. The reversing valve has the effect of reversing the action of the two coils. In the cooling mode, the outdoor coil becomes the condenser coil and the indoor coil the evaporator. In the heating mode, the outdoor coil becomes the evaporator coil and the indoor coil the condenser.

The Heat Pump Cools In the cooling mode, the heat pump works like an air-conditioner or refrigerator. The indoor coil acting as the evaporator coil contains circulating refrigerant that absorbs heat from the inside of the house. It is, of course, colder than the indoor air.

As the refrigerant heats up, it evaporates into a gas. The gas then moves through the pipe to the compressor. In the compressor, the gas is subjected to heavy pressurization, which raises the boiling (vapor) point high enough to change it back into a liquid. It passes out into the outdoor coil, the condenser coil, where it condenses back into a liquid.

This process causes it to give off heat that dissipates into the outside air. Even in summer, this air is colder than the gas itself.

The refrigerant then returns to the indoor coil to pick up more heat, after which the process is repeated.

The effect of the heat pump's action has been to move the heat inside the house to the outside air, making the house cooler and more comfortable.

The Heat Pump Warms In the heating mode, the heat pump does exactly the opposite. The reversing valve causes the outdoor coil to function as the evaporator, and the indoor coil as the

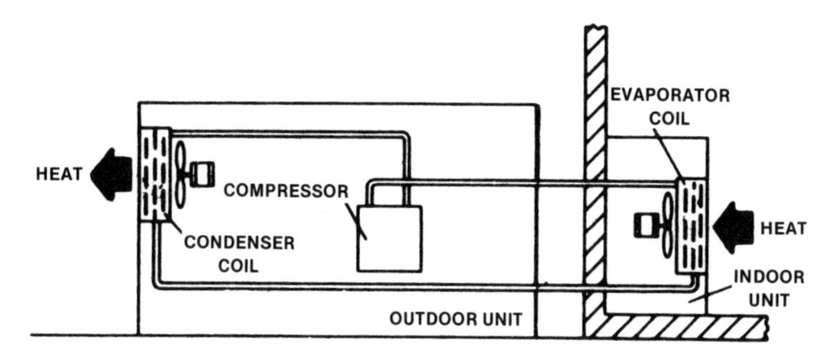

WESTINGHOUSE ELECTRIC CORPORATION

During cooling mode, heat pump sucks in heated air from indoors (right), compresses gas in unit outdoors, turning it to a fluid. As it condenses it releases heat to outdoors (left). Returning fluid cools indoor air.

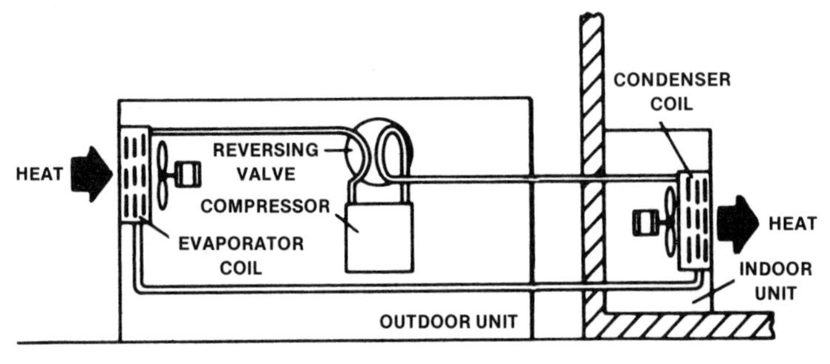

WESTINGHOUSE ELECTRIC CORPORATION

During heating mode, heat pump circulates refrigerant through outdoors cold. Refrigerant, colder than air, is warmed and vaporized until it enters compressor. High-pressured gas flows to indoor coil, now a condenser, where gas turns to liquid. Condensations frees heat which spreads through house.

condenser. As long as the temperature of the refrigerant at the outdoor coil surface is colder than the outside air, heat will flow from the air to the refrigerant.

This transfer of heat causes the refrigerant to warm up and vaporize (boil). The gas then enters the compressor. Here high pressure increases its temperature to a very high degree. The superheated gas flows into the indoor condenser coil, now acting as a low-pressure condenser, where it turns back into a liquid as the boiling point is lowered. This condensation process gives off heat, which spreads through the house with the help of a circulating fan.

The beauty of the heat pump is the fact that it is simply reversed from one season to the next by the simple means of a switch and the action of the reversing valve.

LIMITATIONS OF THE HEAT PUMP

The heat pump does not perform efficiently when the outside temperature drops below 20° F. However, if the heat pump can be made to draw heat from water or air warmed to a modest 50° by flat-plate solar collectors, it then functions with complete efficiency.

That is the reason the combination of the heat pump and the solar collector and storage system makes it a formidable weapon against hot and cold weather.

NEW DEVELOPMENTS IN THE HEAT PUMP

The heat pump business is quite viable today, thanks to the negative effects of the oil and gas crisis on conventional heating systems. Westinghouse has come up with a new design based on subcooling at the condenser rather than superheating at the evaporator. One new element is added to the conventional design: an accumulator-heat exchanger which permits operation from 115° down to −20° F. outdoors.

In the cooling mode, the compressor discharges high-pressure gas to the outdoor coils. The gas gives off heat as it condenses to a liquid. When the liquid leaves the condenser, it is cooled by 10° in the subcooling valve.

Passing through the accumulator-heat exchanger, the liquid is cooled another 45° in addition so that it is very cold by the time it is expanded in the subcooling control valve.

The liquid/vapor goes to the indoor coil where it absorbs heat and returns to the accumulator-heat exchanger once again.

In the accumulator, liquid drops to the bottom and is evaporated by heat exchange with the condensed liquid. The saturated gas in the uppermost section of the exchanger is drawn into the suction U tube and returned to the compressor to complete the cycle.

In the heating mode, the action is reversed.

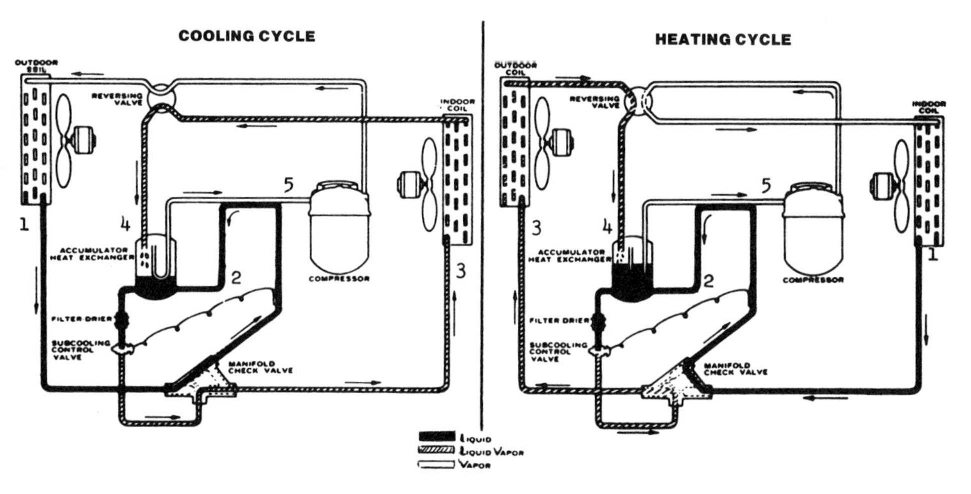

WESTINGHOUSE ELECTRIC CORPORATION

Diagrams illustrate action of heat pump described in accompanying text.

MORE DEVELOPMENTS IN THE HEAT PUMP

Warner's York Division has come out with a redesigned residential heat pump with a built-in computer control system. The company says the system is 20 percent more efficient for heating and 16 percent more efficient for cooling than previous models. The solid-state control system has been engineered to increase the reliability of the unit.

Carrier Corporation and General Electric are also introducing new models of the heat pump.

Because the heat pump is run by electric power, it is usually coupled with an electric-resistance heating unit as a backup when the outside temperature falls too low for safe operation. The addition of solar collectors, bringing the temperature up to 50°, usually means that the system will work without backup.

On the drawing board is a new idea heating scientists are toying with. It's a way of building the capability of storing heat into the heat pump itself.

Tests show that a home with a solar system and a heat pump system combined will use only one-quarter as much electricity for winter heating—or even less—as it will with electric heating of the resistance type.

The only drawback in the solar/heat pump system is that the rooftop collectors sit idle in the summer when the heat pump is reversed to function as a conventional air-conditioner.

SOLAR-BACKED HEAT PUMP SYSTEM

An experiment at the University of Maine at Orono by two graduate students proves that the system of combining heat pump with solar energy is efficient. It works this way.

The heat pump system is composed of compressor, evaporator, expansion valve, and condenser, along with two tanks of water: a cold reservoir of 16,000 gallons and a hot reservoir of 1,600 gallons.

The water in the so-called "cold" reservoir—called cold because it is cooler than the "hot" reservoir—is supplied with low heat from solar collectors, maintaining the cold reservoir at a low temperature. This "cold" water circulates first over the evaporator tubes through which flows a refrigerant, R-500. The refrigerant, at a low temperature and a low atmospheric pressure, is a mixture of liquid and vapor. The "cold" water circulating over the tubes is at a higher temperature. It heats the refrigerant, causing the refrigerant to boil (evaporate) and change to a gas, losing temperature as it does so.

The gas leaves the evaporator colder than it entered. At enforced low pressure in the condenser, the refrigerant changes from liquid-gas to gas.

Next the high pressure of the compressor acts on the gas, greatly increasing its temperature. The hot gas goes into the condenser. Here the pressure is much lower. The temperature of the water flowing through the condenser in tubes is lower than the refrigerant. This causes the refrigerant to condense to a liquid on the outside of the water tubes.

Condensation gives off heat which is transferred to the water, causing it to heat up. The refrigerant, which has entered as a gas, leaves as a liquid. The water in the "hot" reservoir is heated.

The liquid refrigerant enters the expansion valve, which reduces the pressure, lowers the boiling point, and causes some of it to flash into a gas. When this mixture is pulled into the evaporator tubes again, the cycle continues.

A new 25,000-square-foot building at New Mexico State University at Las Cruces has such a system. Its designer plans to incorporate such systems in larger buildings on which he is now working.

5

A Solar Hot Water System

The average home uses about 40 percent of its heating energy keeping the supply of domestic hot water up to temperature. Cutting down on the amount of heating energy, whether it is supplied by oil, by natural gas, or by electricity, can provide considerable savings to a household budget.

As far back as the 1920s and 1930s, before cheap natural gas and electricity became available, there were about 60,000 domestic solar hot water heating systems in the southern and western states. Many of them were in Florida.

Today solar hot water heating is standard in both Israel and Australia. Nobody really knows how many hot water systems are installed in this country, but they number in the many thousands. Most of them are working well and paying for themselves in lower fuel bills.

Estimates are that if 75 percent of the homes in the country used solar hot water systems today, the United States would save 250,000 barrels of oil *each day*.

More homes use solar energy to provide hotwater heating than to provide any other type of heating or cooling. In fact, the only other type of solar heating that even compares to it in widespread usage is solar heating for swimming pools.

THE TYPICAL HOT WATER SYSTEM

The hot water heating system differs slightly from the pool water system. Each will be described separately. Simply stated, solar collectors for hot water systems are constructed to provide a higher degree of heat than collectors for pool heating systems.

Yet solar installation of hot water systems is a simple matter. Commercial units are available in most localities. There are units that come in one piece and can be added to the home heating plant with minimum expertise. Others are more complicated and must be installed by professionals with expert knowledge in solar work.

The beauty of the hot water system is that in most cases the homeowner does not have to make major construction or plumbing changes in his house or provide room for a lot of new equipment.

Usually the only changes are the addition of solar collectors, a storage tank, and the plumbing necessary to carry the heated water from the collector to the storage zone. In rare cases, even the storage tank can be eliminated, with the solar heat transferred directly to the hot water heater.

The only fittings needed in addition to solar collectors, plumbing, and storage tank are the proper kinds of pumps, controls, and thermostats.

An effective system can provide 80 percent of the household's demand for hot water. In optimum conditions, it can even provide more. Naturally, in each case, a backup water heater is needed for days when there is not enough sunshine to supply the storage tank with heat sufficient for service.

The first consideration in equipping a house with a solar hot water system is the size of the family that will use it. The average adult in the United States uses between 15 and 20 gallons of hot water each day. That means that the total usage per day for a family of four varies between 60 and 80 gallons.

That doesn't mean that an 80-gallon storage tank is needed. But it does mean that the total capacity of both hot water tank and storage tank should be about 80 gallons.

An average home hot water system can operate with a 30-gallon gas or electric water heater, coupled with a 50-gallon solar storage tank. The 50-gallon storage tank holds the heat gathered by the

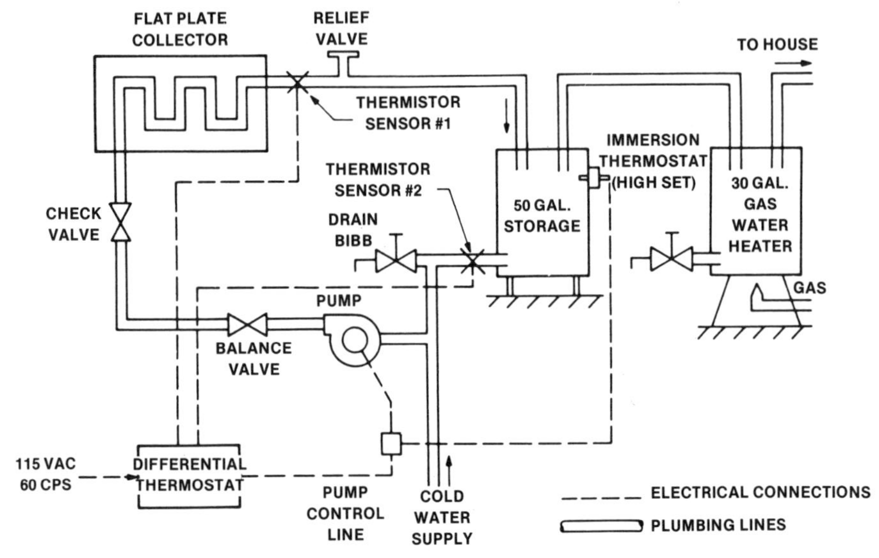

RHO SIGMA INCORPORATED

Schematic drawing shows sensors, valves, and pumps installed in typical domestic hot water solar heating system.

solar collectors. The 30-gallon hot water heater draws on the storage supply until it is exhausted and then provides the backup through electric power or gas.

HOW MUCH SOLAR COLLECTOR AREA?

That brings up the question of the amount of solar collector area necessary for supplying a 50-gallon storage tank with proper heat. Experts figure that the collection area for a domestic water heater in an average residence should be about one square foot for every gallon of hot water required per day.

In certain localities where there is a high percentage of cloud cover, one and one-half or even two square feet of collection area may be required for each gallon. The charts in Chapter Fourteen give some indication of the variation in collector areas.

Most solar collectors for hot water systems deliver heat of at least 140° F. or over to the storage tank. Sensors control the flow of heated water to the storage zone to prevent delivery of water warmed to a temperature less than that of the storage tank. In that event, the storage tank would then give up heat to the collector fluid.

GRUMMAN CORPORATION
Two fluid-medium solar collectors for domestic hot water supply are contained easily on sloping roof of Long Island, N.Y., home.

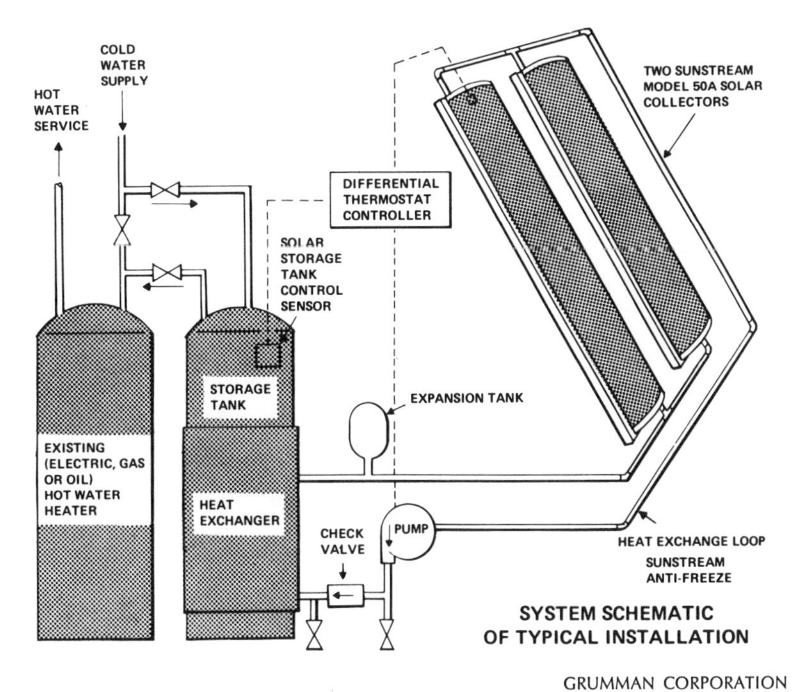

GRUMMAN CORPORATION
Fluid medium, containing antifreeze, imparts heat to storage tank through heat exchanger. House draws water from conventional tank.

Most hot water heaters are set at 140° F. However, during the recent energy crisis experts discovered that such a temperature is not really necessary for most household needs. Many hot water heaters are now set at 135° F. or even lower. In some localities water at 120° F. is considered plenty warm for hot water use.

It is a good idea to investigate your own hot water system. You may be setting the thermostat too high for your needs. In other words, you may be paying for water that is hotter than you want. A difference of 5° in the hot water thermostat may save you up to fifty dollars a year in heating bills.

WILL SOLAR HEAT COLLECTION CUT HOT WATER BILLS?

For the house which heats its water by electric-resistance heat, the use of solar energy even to a partial extent will cut the size of electricity bills considerably. Within a short time the cost of installation will be recovered.

For the house which heats its water with natural gas, the installation of solar equipment will probably take some time to pay for, even if natural gas prices don't maintain their present level.

For an oil-heated water system, the installation of solar equipment will save some money in the oil bill with oil prices certain to continue up. It will take less time to equate savings with the cost of installation.

Further discussion on financing of solar heating and cooling systems is contained in Chapter Fifteen.

A TYPICAL SOLAR HOT WATER SYSTEM

To equip the average-sized house with four people living in it with a solar collector system to operate a domestic hot water system using a 30-gallon gas or electric water heater, the following additions would normally need to be made:

A 50-gallon storage tank should be installed near the hot water heater to hold and store solar heat.

Several solar collectors should be installed on the roof of the house, or somewhere nearby—even in the yard—to collect the sun's heat. The number of square feet of the collector area should be about one square foot to each gallon of the storage tank. Sometimes other considerations, like geographic differences, should be taken into account.

Pumps, sensors, and valves should be installed along with a system of pipe to carry and control the flow of heated water from collector plates to the storage zone.

In the typical home installation, the 50-gallon storage tank operates as a preheat system for the conventional 30-gallon water heater.

A 50-gallon storage tank needs a solar collector area of about 40 or 50 square feet. At average efficiency, the system should supply about 60 percent or more of the hot water needed for a family of four under generally sunny conditions.

During extended periods of overcast and storms, the 30-gallon gas or electric system would function as backup to the solar system.

During clear sunny days, even under low temperature conditions, the storage tank water would rise to very high temperatures—180° to 200° F. To prevent scalding to those using the water, a thermostat control set at about 160° F. should be used in the storage tank to keep out solar water that is too hot for the storage tank to handle. Without such a thermostat setting, temperatures could exceed 180° on a clear sunny day.

COMMERCIAL SOLAR HOT WATER UNITS

There is fairly widespread use of solar hot water heating units. Several companies manufacture units and parts for sale and installation.

Study of a few of them will indicate the kind of design and type of service each provides. Some are complicated systems, and others are one-piece affairs which combine solar collector and circulating pump in one unit.

ONE-PIECE UNIT FOR HOT WATER HEATER

Sol-Heet is a one-unit system. The heart of this system is the solar collector, which is mounted anywhere it can catch the sunlight. This site is usually the roof of the house.

Within it the collector also contains the unit's complete control system, including a circulating pump. A glass-topped absorber box, it weighs approximately 120 pounds. It is designed to be mounted on the roof of a house, on a carport, or on ground level.

Once mounted, the collector is attached to pipes leading to the household's existing hot water system.

In a model environment of continual sunshine, this simple one-piece system will provide an endless supply of hot water. However, because the number of cloudy days varies from one place to another, the unit must be attached to an already functioning hot water system that can take over when the sun fails to shine for extended periods.

Once the unit is installed, it becomes the primary source of heat for the hot water system, and the gas or electric heater becomes a backup in case of bad weather.

With the hot water tank set at 120° F. this system can save up to 35 percent of household electrical or gas energy.

A CYLINDRICAL SOLAR COLLECTOR

Another type of solar collector commercially available for use with a hot water system is the Sav, manufactured by Fred Rice Productions, Inc. The collector is constructed in the shape of a large cylinder, and its structure makes it act both as heat collector and storage tank.

Air spaces around the interior glass cylinder provide for both heating and insulation at the same time. There are two separate glass compartments—one called the inner glass house, and the other the outer glass house.

The heated-water outlet of the Sav unit is connected by pipe to the hot water heater, and the cold water inlet is connected to a water tank placed between collector and hot water tank. This tank supplies cold water from the house's pipes for circulation through the collector.

When the water warms up in the collector, it moves upward through the system, becoming heated as it does so, to the hot water heater. There it provides hot water usually at about 140° to 160° F.

The collector units are carefully calibrated. Each unit contains about 12 gallons of water. When heated, one unit adds 12 gallons

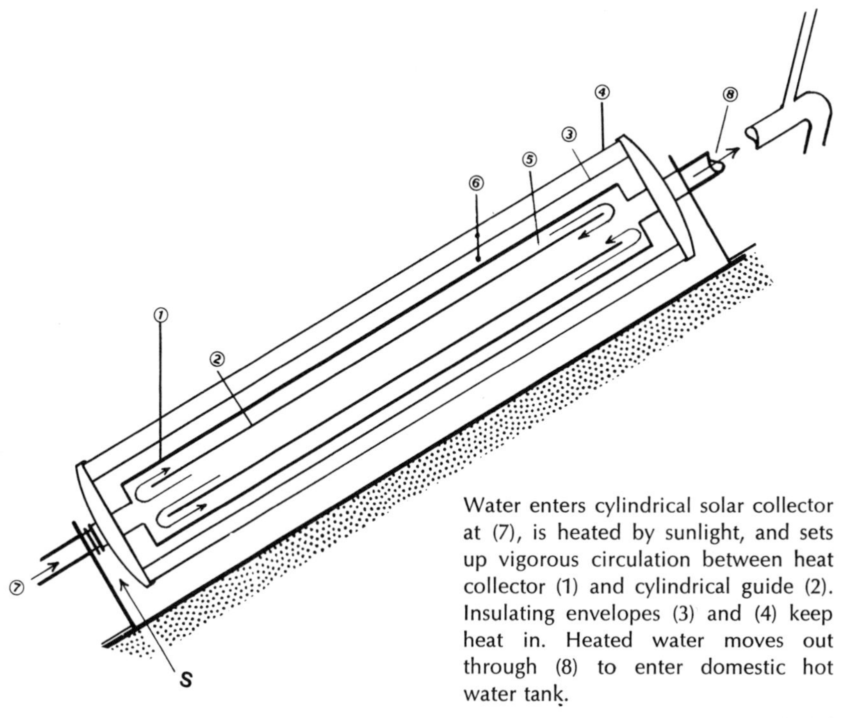

Water enters cylindrical solar collector at (7), is heated by sunlight, and sets up vigorous circulation between heat collector (1) and cylindrical guide (2). Insulating envelopes (3) and (4) keep heat in. Heated water moves out through (8) to enter domestic hot water tank.

FRED RICE PRODUCTIONS, INC.

of hot water to the domestic hot water tank. Two units will provide 24 gallons, and so on.

Each Sav unit is about four feet long, and weighs 38 pounds. With the collector absorbing sunlight in a horizontal position, the unit provides a temperature rise of up to 46 degrees per hour. If tilted to the correct angle for the proper geographical latitude, the temperature rise will be increased.

The unit needs no pumps of thermostats. It requires only connection to the existing hot water heater. That system can be set to any desired heat up to 160° F.

SYSTEM WITH HEAT EXCHANGER

Revere's system for providing domestic hot water service is a little more complicated, containing as it does collectors, pumps, storage tank, differential thermostat, heat exchanger, expansion tank, plus air vents and an assortment of relief, balancing, and control valves.

The solar collector contains a heat absorber plate made of carrier tubes integrated into the sheet of copper. Painted black, the absorber plate is mounted in an aluminum box covered with glass layers. The box is well insulated at the back.

Storage tanks come in sizes from 66 gallons to 120 gallons. A heat-exchange system built into the storage tank allows a fluid medium to recirculate from collector to tank. The fluid medium is antifreeze and water, a mixture which prevents freeze-up of fluid on cold nights.

The system works this way:

When the fluid in the solar collector is more than 15° higher than the water in the storage tank, the pump starts up, delivering the heated fluid to the heat exchanger in the storage tank. Heat from the fluid warms the water around the exchanger.

When the temperature difference drops to less than 5°, the thermostat stops the pump, since not much heat is exchanged at a difference of only 5°.

FROM HOT AIR TO HOT WATER

Solaron Corporation's air-medium hot water heating system also employs a heat exchanger. In this model, the heat exchanger does not transfer heat from fluid to fluid, but from air to fluid.

The system uses two solar collectors—sometimes four in larger residences or buildings—which heat the air circulating through the panels.

Hot air is ducted out to the heat-exchange unit, where the heat is transferred from the air medium to water pumped in from the domestic water system.

This solar-heated water circulates into the storage tank as long as the heated water is warmer than that in the tank. When it reaches 160° F., a sensor stops the pump.

The storage tank is connected to a regular domestic hot water tank. Water for use in the house is drawn from the conventional tank, and is replaced simultaneously by water from the storage tank.

If water in the storage tank falls below 120° F., or whatever the thermostat is set for, the regular domestic system supplies heat for the water to bring it up to temperature.

This air-medium hot water system can also be operated with

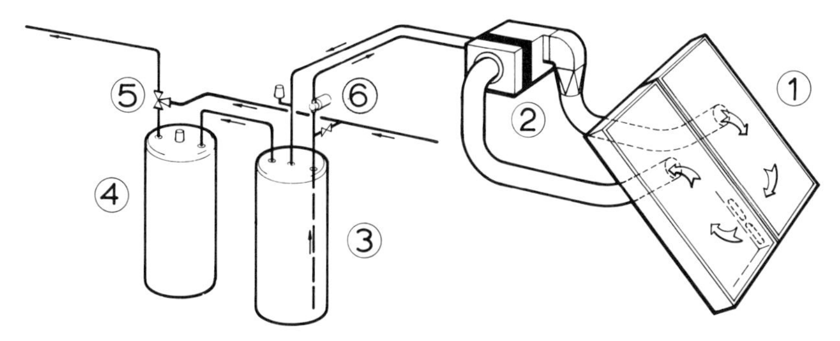

SOLARON CORPORATION
Diagram shows air-medium solar collector system used with domestic hot water system. Hot air in collectors (1) moves to heat exchanger unit (2), where it imparts heat to water circulating from storage tank (3) and moved by pump (6). When hot water is drawn from regular hot water tank (4), storage tank water replenishes supply. Mixing valve (5) adds cold water if temperature is over 120°.

only one tank. In that setup, the tank serves both as storage tank and as domestic hot water tank.

THE CURVED-TOP HOT WATER COLLECTOR

Grumman Corporation's fluid-medium hot water heating system is designed for use in almost any climate. Called Sunstream, it uses an antifreeze solution in order to prevent collectors from freezing in cold winter weather.

Each collector has a distinctive cover made of curved acrylic designed for a pleasing appearance, light weight, and extra strength against breakage. The curved surface also enables rain and wind to keep the surface clean of dust and drifting debris.

It works this way: Antifreeze solution enters the bottom of the collector, heats up in the sun's rays, then moves on to the heat exchanger, where it gives up its heat to the water in the storage tank.

The storage tank is connected to the house water supply. When water is drawn out of the conventional water heater, the storage tank replenisehes the supply in the hot water tank.

Water from the house's main supply enters the storage tank, to be heated in turn by solar-warmed water out of the collectors. The system can also be operated with only one storage tank serving as both storage and supply tank for the house, provided the tank is big enough.

In actuality, the solar heating system functions as a preheater to the existing system.

TYPICAL INSTALLATIONS OF HOT WATER SYSTEMS

The solar hot water system can be added to existing homes without great difficulty. Installing a solar system on a house that is already built is called "retrofitting."

Retrofit System in Vermont Farmhouse The unique feature about the hot water system adapted in a 200-year-old Vermont farmhouse that was once the home of U.S. President Chester A. Arthur is the successful installation of solar collectors that do *not* face south.

In fact, they face both east *and* west.

The portion of the house on which the solar collectors are fitted is an extension of the original structure. The Hinesburg home of President Arthur is oriented the wrong way for south-facing application of solar collectors.

That didn't bother the owner, who has a Ph.D. in engineering. He calculated that if it was impossible to install solar collectors to the south, he could catch at least a portion of the sun's energy by putting five collectors to the east, and four more to the west. In that way, he could catch the morning's energy first and the afternoon's energy last as the sun moved across the sky.

The system is simplicity itself. Collector panels absorb the sun's heat as the fluid is pumped through the panels from top to bottom. The heated fluid travels to a heat exchanger in a hot water storage tank, where the fluid gives up its heat to that in the tank. Another heat exchanger carries heat from the storage tank to the domestic hot water tank. There the hot water is ready to be drawn whenever wanted.

The collector panels are mounted on top of an ordinary shingle roof. No change in structure or slope has been necessary.

The system provides 80 percent of the household's yearly demand for hot water. The domestic hot water heater was in the house before retrofitting. Hot water leaves the taps at 140° F.

ERDA
Multifamily dwellings in El Toro, California, get 70 percent of domestic hot water from flat-plate fluid-medium copper solar collectors.

When cloudy days shut off the sun's energy, a backup heater provides hot water for the household.

Hot Water System with Heat Exchange In Vienna, Virginia, a solar system that provides heat for domestic hot water use features a heat-exchange system inside the hot water tank. The installation has no need for a secondary storage tank.

The solar collectors are mounted on the south-facing slope of a pitched roof, with the collector boxes tilted at about a 38° angle—approximately the north latitude of the Vienna area (about 38° 55′ N).

Three liquid-medium flat-plate collectors manufactured by Sunworks are used in the system. The collection area totals 21 square feet.

The 65-gallon hot water tank in the house is a conventional water heater with a heat-exchange system added to it. The tank itself is an electric-operated water heater with a 4.5 kilowatt heating element. When solar energy fails to heat the water enough, the electrical system and heating element take over.

The system works very simply. When the sun heats the solar

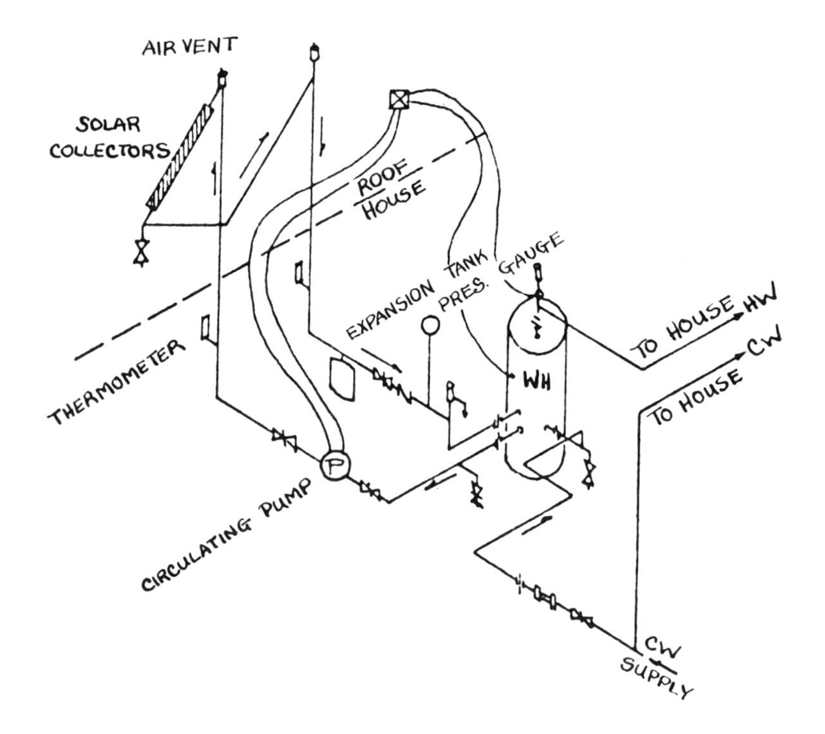

HUD

Hot water supply comes from three collectors on sloping roof of house. Collectors are tilted and space covered to prevent wind damage. Diagram shows how fluid in solar collector warms water in heater through heat exchanger and then recirculates it to roof.

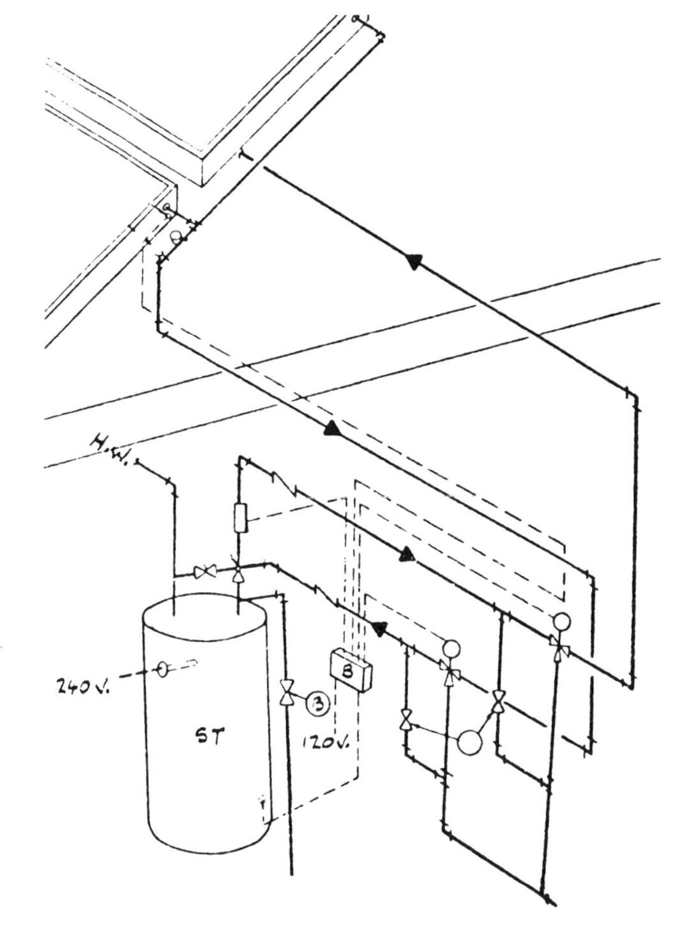

HUD

Solar collectors get special mounting on flat-roof multifamily dwelling. Sun heats domestic hot water for eight families. Diagram shows hot water from collectors flows directly into hot water tank.

collectors, a circulating pump moves the heated liquid in the solar collectors down to the water heater. The heated fluid travels through a coil inside the water heater, giving up heat to the water in the tank.

An expansion tank is attached to the system to allow excess water to leak into it rather than create excessively high pressure in the system.

Sensors control the action of the circulating pump. Both the solar collector circuit and the hot water tank supply are tapped from the house water supply.

Solar Hot Water for Multifamily Dwelling An eight-family garden apartment building in St. Petersburg, Florida, gets its domestic hot water from a solar collector system that has been integrated into a regular water system, with very little added equipment.

The area of the solar collectors totals 144 square feet. The collectors, by Gulf Thermal Corporation, are liquid-cooled flat-plates with an automatic drain-down system to prevent freezing in cold weather. They are mounted at an angle of about 28°, the latitude of the St. Petersburg area.

Storage capacity for the collectors totals 320 gallons, with four 80-gallon storage tanks used to service the eight apartments. The water that goes to the collectors is tapped originally from the domestic water supply. It is then pumped to the collector area and brought back down into the hot water storage tank when it is sufficiently hot. From there it is drawn directly from the storage tank by the residents.

The four water heaters contain electric heating elements and can provide hot water if the solar system fails to produce sufficient temperatures during cloudy or overcast weather.

Hot water is distributed to the eight apartments from the storage tanks by a conventional plumbing network.

The pipe system connecting the collector area and the water heater storage tanks is carefully insulated to reduce heat loss from the collector to the storage area.

Solar heat is used only for the supply of domestic hot water for those living in the apartment complex.

6

A Solar Swimming Pool

According to fairly recent and reliable figures, there are about 10,000 swimming pools heated by solar energy in the United States. Interest in this facet of solar heating is catching on fast. Three out of four of these systems have been installed within the past five years.

A large percentage of these solar-heated pools are located, as might be expected, in California or in Arizona. One reason for this is a recent ban on the use of natural gas for pool heating in some western states. Gas was once used exclusively to heat these pools. The shortage of natural gas brought to light during the recent 1976–77 winter may cause other states to ban gas for pools and cause pool owners to install solar systems.

California and Arizona do not contain all the solar-heated swimming pools, by any means. There are great numbers in Florida, in other southern states, and quite a few in the North as well. Many of these latter are enclosed pools.

Now people all over the country are turning to the use of solar energy to heat their pools in order to extend their swimming season into the spring and fall.

HOW A SOLAR POOL HEATER WORKS

A system for heating a swimming pool by means of solar energy differs slightly from the type used to heat a domestic hot water system. In certain instances, a different type of solar collector may be used, although many pools use conventional solar collectors or even make use of a space heating system during the months when solar energy is available in quantity.

The difference in the two systems is in heat capacity. A solar hot water heater operates on the principle of providing heat through solar collectors to warm a storage tank and/or hot water tank heated to 120° F. or more. In some cases, the surface temperature of the solar collector in operation is heated to 200° F. or more, even as high as 300° F.

The principle of the system used to warm a swimming pool is somewhat different. The heat collected from the sun does not have to be quite so hot as that which warms the hot water heater. The pool heater is actually trying to warm a *large* quantity of water only several degrees, whereas the hot water heater collector is trying to warm a *small* quantity of water to a high degree of heat.

Solar collectors for heating home pools are usually mounted separately, but they can be part of entire heating and cooling system. GRUMMAN CORPORATION

There are two ways of accomplishing this. One is to heat a small quantity of water to a very high temperature—140° F. or so—and let the heat spread through the whole pool. The other way is to circulate a large volume of water quickly through a solar collector, thereby keeping the collector cool, and warming all the water by several degrees. Because of this method of operation, the solar collector exclusively for pool heating can work at a low temperature rather than a high one.

Pool Heating Solar Collector For example, a pool heater can deliver a large volume of water only 2° to 5° warmer than it was when it entered the collector. The small heat rise more than makes up in volume what it sacrifices in higher temperatures. Keeping a large body of water at about 80° F. is actually easier than keeping a small portion of water at 140° F.

High-temperature solar collectors must be made up of metal absorber plates to withstand the high temperatures generated by the "greenhouse" effect of the glass cover. High-performance solar collectors with black, selective surface coating and multilevel glass covering sometimes operate at temperatures well above 200° F.

Since a solar pool collector does not have to withstand such high temperatures, it can be constructed of a different type of material. It does not, actually, have to be covered with two panes of glass or clear plastic, nor does it have to be so carefully insulated. A typical solar collector for a swimming pool heating system can be made of plastic.

Because a solar collector for pool heating does not have a glass or plastic cover and is not insulated, it can sit out in the sun without being heated to high temperatures that might damage it.

The main advantage of using plastic in the solar collector is its price. Pool heating can be very inexpensive, compared to hot water and space heating.

INSTALLATION OF A POOL HEATING SYSTEM

The installation and planning of pool heating is also a great deal less complicated than space heating and even than hot water heating. The question of heat storage, circulation, and control system can be forgotten.

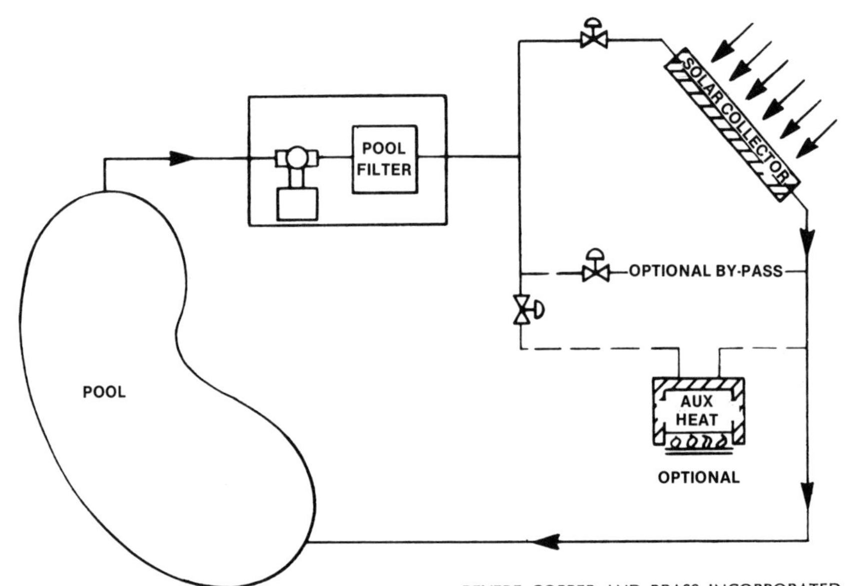

REVERE COPPER AND BRASS INCORPORATED

Diagram shows typical swimming pool heating system, utilizing conventional solar collectors. Backup system is unnecessary except in very bad weather.

The swimming pool acts as its own storage zone and circulating system. It even acts as a kind of pump system, with the pool's circulation/filtration system providing an ideal flow of water through the collectors.

SIZE, INCLINATION, AND SITING OF COLLECTORS

The recommended area for solar heat collection for a swimming pool is a minimum of one-half to three-quarters of the pool's surface area.

A pool with an area of 800 square feet needs from 400 to 600 square feet of collector panel area.

The best orientation for a pool-heating collector panel for year-round heating is an inclination equal to the latitude of the installation, so that the sun will shine squarely on the collector.

For summer heating only, the panels should be inclined at an angle equal to the latitude of the installation minus 10° to 15°. For winter heating only, the panels should be inclined at an angle equal to the latitude plus 10° to 15°.

The ideal site for the location of the panels is a south-facing pitched roof, as close to the pool as possible. However, a western exposure is acceptable if the panel area is increased slightly. An eastern exposure tends to deliver about 50 percent efficiency, and is only marginally economical. A horizontal exposure on a flat roof tends to deliver about two-thirds efficiency; increasing the collector area to three-quarters will improve the efficiency.

A northern exposure is unacceptable.

COMMERCIALLY AVAILABLE SYSTEMS

There are an endless number and variety of components available for solar pool heating systems. The heart of each system is the collector. In some cases the entire system is locked up in the collector and its connections with the pool. This provides ease of installation and maintenance.

One typical collector designed especially for heating a swimming pool is the Fafco solar panel. The panel is made strictly of plastic, with pipe tubes at top and bottom for input and output of swimming pool water.

Through the middle of the solar panel run hundreds of tiny parallel channels which allow the water from the bottom of the panel to move up to the top of the panel, passing through virtually every square foot of the panel exposed to the sun.

Unlike the typical hot water solar panel with its components, this pool heater panel is all in one piece. The plastic panel acts as absorber plate, conduit system, and heat trap all in one. Special chemicals toughen the plastic material to make it more resistant

Lightweight, easy-to-install, all-plastic solar collector can keep swimming pool at comfortable temperature.

FAFCO INCORPORATED

to the sun's ultraviolet rays, which tend to break down certain plastic formulations. Ultraviolet rays cause discoloration, brittling, and cracking in untreated plastic.

Chemical additives to this plastic resin make the resultant material an effective heat absorber. Carbon black, used as a stabilizer, screens out solar rays all over the plastic surface. It turns the plastic opaque to both visible and ultraviolet light, preventing ultraviolet radiation from penetrating beyond the surface of the collector.

Carbon black also toughens the polyolefin resin in the absorber plate so that the material will last for twenty years or more without discoloration or breakdown. Carbon black improves the weather resistance of the plastic.

The panels come 51⅜ inches wide, in lengths of 8 and 10 feet. They can be mounted on the roof itself or can be put up in custom-built racks. A 10-foot panel weighs 16 pounds. When full of water it weighs only 58 pounds, a little over a pound a square foot.

Each panel is capable of delivering water at the rate of 8 gallons per minute, the recommended flow.

For year-round operation, a supplementary heater is required with this collector for most parts of the country. In Florida, the

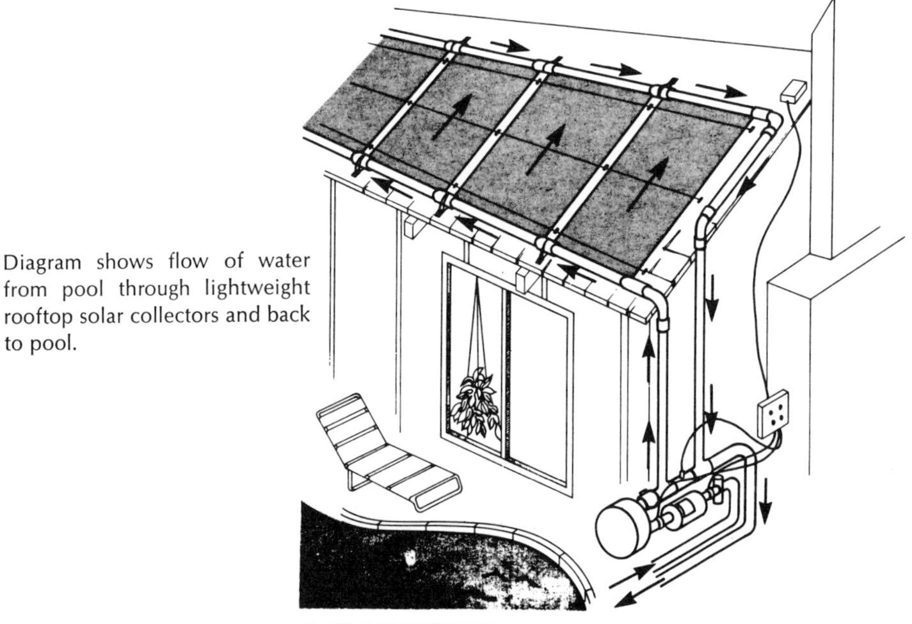

Diagram shows flow of water from pool through lightweight rooftop solar collectors and back to pool.

FAFCO INCORPORATED

system will provide winter performance, if from three-quarters to all the pool's area is duplicated in solar panels.

PLASTIC SOLAR COLLECTOR FOR POOL HEATING

Another solar collector exclusively designed for pool heating is manufactured by the Sundo Company. This unit is a one-piece plastic absorber/conduit/heat trap through which water runs from and back to the pool.

The panel is made of ABS, acrylonitrile-butadiene-styrene, a plastic that has high heat resistance and chemical resistance. The material is impregnated with carbon black, making the entire panel heat-absorbent.

Water flows from bottom to top through rectangular channels 6 inches wide and 5/16-inch thick. Each panel has a number of these 6-inch channels side by side.

For best results, the panels should be placed on a south-facing surface, although a southeast or southwest exposure may be used with only some loss in efficiency.

The hydraulics in the system will usually operate by means of the pool pump, although if the water has to be pumped to a height of 8 feet or more for a roof installation, a motor of greater horsepower may be needed.

Units are lightweight and should not adversely affect any roof structure they are mounted on. When filled with water, each unit weighs approximately 1 pound per square foot.

The panels are made in 4-by-8- and 4-by-10-foot lengths.

Installed in a residential pool in La Habra, California, the system delivered the following temperatures in random months:

In May, when the air temperature varied from 80° F. to 56° F., the pool remained at 84° F. In June, with the air 73° to 68°, the pool was 78°. In July, with the air from 91° to 68°, the pool was 84°. In October, with the air 97° to 63°, the pool was at 79°. In November, with the air 77° to 63°, the pool was 72°.

DO-IT-YOURSELF POOL HEATER

Although it is not yet the easiest thing in the world, it is possible for the experienced do-it-yourselfer to put together a viable solar pool heating system himself.

In Los Angeles, California, a manufacturer's representative for the plumbing, heating, and cooling industries built a solar heater for his 40,000-gallon swimming pool in the Hollywood Hills area.

Working alone, he built the pool heater in about two weeks' leisure time. He spent no more than $1,000 on materials, including Type M copper tube, 16-ounce copper sheet, weatherproof board, and pipes to carry the water to and from the collector area. The design for the collectors appears in Copper Development Association's forty-six-page booklet, "How to Design and Build a Solar Swimming Pool Heater."

He made his collectors in 4-by-8 modules. Water flowing from the pool through the copper tubes absorbs heat from the copper backing and carries it back into the pool, heated.

Depending on the pool size and what part of the country it is located in, this do-it-yourselfer believes that a pool owner can save from $50 to $100 a month on energy bills by using solar heating for the pool.

For the skilled do-it-yourselfer, the construction of solar collectors for use in heating a swimming pool is a snap. Just follow directions in a do-it-yourself book. COPPER DEVELOPMENT ASSOCIATION

AN ENCLOSED POOL HEATED BY THE SUN

All solar-heated pools aren't necessarily exposed to the open air. Because of the weather in the New England area, many pools that far north have been put in enclosures.

These enclosed pool heating systems can be elaborate affairs. In fact, in the past few years, while about 7,500 fairly simple solar energy systems have been installed throughout the country to heat swimming pools, about a hundred enclosed solar swimming pool heaters have been installed in the vicinity of metropolitan New York, despite the cost of the installations. In most cases, a backup heat source is needed to provide heat if the winter closes in unexpectedly soon.

Most of these installations utilize conventional solar heating collectors used for space heat and hot water heat, and not warm-weather pool heating panels. In a typical installation in Connecticut, using a Sunworks solar heater, the enclosed pool is located in a 40-by-24-foot wing of a house. It is linked to the main house by sliding glass doors.

The solar collectors include five copper-plate panels mounted on the south roof of the building. The panels are mounted at a 37° angle. Each is a 3-by-7-foot panel, consisting of a black-painted copper absorber with a double glass cover.

The plate is a fluid-medium collector, containing copper tubes in typical S shape attached to black-painted absorber plate.

An electric pump circulates the fluid medium—a mixture of water and antifreeze—through the copper tubes when a thermostat control decides that the shining sun is hot enough to heat the fluid medium in the collector at about 100° to 110° F.

The warmed collector fluid is then pumped down through the pipes to a heat exchanger. The pool's filter system, consisting of two coiled pipes laid closely together, is connected to the heat exchanger. The newly warmed water in the heat exchanger flows into the cooler pool water, raising its temperature by 6° to 12° F.

The system is designed to maintain the pool water at a comfortable 75° temperature. In order to keep the pool's heat from radiating out into the cooler air, the pool enclosure is well insulated, and is also heated by an oil-fired forced-air furnace located in the main house.

Thermopane sliding glass doors are used throughout the structure, as is fiberglass insulation—3½ inches thick in the walls and 7 inches thick in the roof.

Because excessive moisture is apt to build up in any pool house, the designer has installed four Plexiglas air vents in the roof of the building. These can be opened when the humidity level gets too high and moisture begins forming on the inside walls and glass.

Hardware and installation cost about $2,500 for the solar heater system. That is about four times what a propane gas or natural gas heater costs.

However, there is one advantage to the solar heater system: it can be hooked into the house's hot water supply during the summer when pool temperatures are easy to maintain.

A valve in the solar heating system can be manually operated to redirect the solar-heated water through a tank. There, with the aid of an auxiliary electric element, it helps warm the house's domestic hot water supply.

The heater is so efficient in the summer that the oil-fired water-heated boiler can be shut down all during summer weather.

Throughout the year, the temperature of the pool can be maintained at about 74° to 76° F. In winter, after a cloudy spell of three or four days, the pool temperature may drop to 68°, but it has never gone lower than that.

As soon as the sun comes out again, the collectors begin functioning and reheat the pool almost immediately.

COMPLETE SOLAR HOME POOL HEATER

In a completely heated and cooled solar home, the heating of the pool is sometimes included in the entire system. Usually, however, the pool system will be separate from the rest of the house.

A specially designed ranch-style house in Coral Springs near Fort Lauderdale, Florida, is testing out a solar heating and cooling system which features a heat pump developed by Westinghouse Electric. A solar-heated swimming pool is included in the residence.

The house is described more fully in Chapter Nine.

Each separate type of heat—space heating, hot water heating, and swimming pool heating—operates separately, although all are tied in to the overall system.

WESTINGHOUSE ELECTRIC CORPORATION
Nine black plastic panels mounted on the ground heat swimming pool of Florida solar home. Plastic surface is cool enough to touch. The pool is enclosed for year-round use.

Heat for the swimming pool of the Coral Springs house comes from the energy collected by nine black plastic panels mounted on the ground near the pool. Normally these panels would be mounted on the roof, slanted at the correct angle. They are installed at ground level for demonstration purposes.

The black plastic surface remains cool enough to touch safely, even in hot sunlight. Inlet and outlet water temperatures can be read on two thermometers.

Year-round use, use even during the coldest of the winter months, is possible because of the enclosure which surrounds the entire pool area.

COPPER COLLECTORS FOR POOL HEATING

An experimental house constructed in Tucson, Arizona, provides solar heat for an open-air swimming pool in addition to providing heating and cooling capability for the entire house.

The Decade 80 Solar House, developed by Copper Development Association, is a large, rambling structure containing about 3,000 square feet of living area. A one-story, three-bedroom house, it has a pool attached and a small guest wing. It is described more completely in Chapter Eight.

Essentially 100 percent of the heating power for the house comes from the sun, and 70 percent of its cooling energy.

The swimming pool is heated by a separate installation of solar heating panels. These collector boxes are mounted on a separate guest wing of the house. Sloping at about 40°, the guest-wing roof favors swimming pool heating during spring and fall.

Although of the same construction as the solar collector panels used to provide energy for heating and cooling, the panels supplying the swimming pool with heat are not covered with glass.

The pool heating system does not operate completely independently of the rest of the house. The air-conditioning system used to cool the house is an absorption type, explained later in Chapter Eight. A condenser for the absorption machine is attached to a cooling tower on the house roof. Heat from that system is added to the pool's collector fluid through a heat exchanger near the air-conditioner's cooling tower. The idea is to make maximum use of all heat possible in the integrated system.

COPPER DEVELOPMENT ASSOCIATION
Heat for this kidney-shaped pool in Arizona is supplied by fluid-medium copper solar collectors mounted on roof nearby.

A SIMPLE DO-IT-YOURSELF TIP

If you don't have the money to spend on a pool heating system, or if you don't have enough building savvy to follow the intricate instructions in do-it-yourself booklets, you can always make a simple pool heater.

Coil about 2,500 feet of one-half-inch black plastic pipe on top of your garage or house. Hook up the hose to the swimming pool, and then cut in the pool's filter pump. Pump the pool water through the pipe on the roof, picking up heat in the process, and send it back into the pool. The amount of heat the coil picks up will surprise you.

The device should keep the temperature in the water in the 80s.

At night, cover the pool with a plastic top to cut down the heat loss into the cool night air.

7

Solar Space Heating

About one-fourth of the world's consumption of fuel is spent in heating the interiors of buildings—including public buildings, factories, and residences. The process of heating interiors is called space heating.

The technique for supplying space heating by means of solar energy is becoming better known each day to architects and engineers. Solar space heating systems are being installed gradually in new homes as well as in old.

Financial considerations are responsible for keeping the number of solar space heating systems to a minimum. It is simply cheaper to heat with gas- or oil-fueled or electrical heating systems.

Hot water heaters and pool heaters, described in Chapters Five and Six, are simple to install in an already existing house. Space heaters are not easy to install, and in many cases cause such a disruption of the existing home that their addition or substitution becomes impractical.

The truth of the matter is that most houses today are not sited, located, designed, or built with solar space heating in mind. Such systems demand a large area for solar collectors that must be integrated into the design and construction of the house itself.

Siting. Heavily wooded areas which obscure the sun are impractical for the planning of a solar home, unless large clearings are made in the timber—so is a building site where existing structures over one story high can cut out the sunlight even partially.

Location. Heating a home by means of solar energy requires a great deal of sunshine in the particular geographical area of the country where the house is located. In localities where there are a lot of overcast and cloudy days, a solar space heating system cannot be expected to deliver as much as it will in a cloudless and sunny climate.

This does not actually mean that there is anyplace in the country where a solar heating system cannot be used. But it does mean that a backup furnace will be necessary in most areas. (See Chapter Fourteen.)

Construction. A badly insulated home is an impediment for a solar space heating system. A tremendous amount of carefully captured solar energy can be lost through leaks and holes. Solar homes must be tightly insulated. The average house is not insulated very carefully. Up to now, the cheapness of fossil fuel made extensive insulation unnecessary.

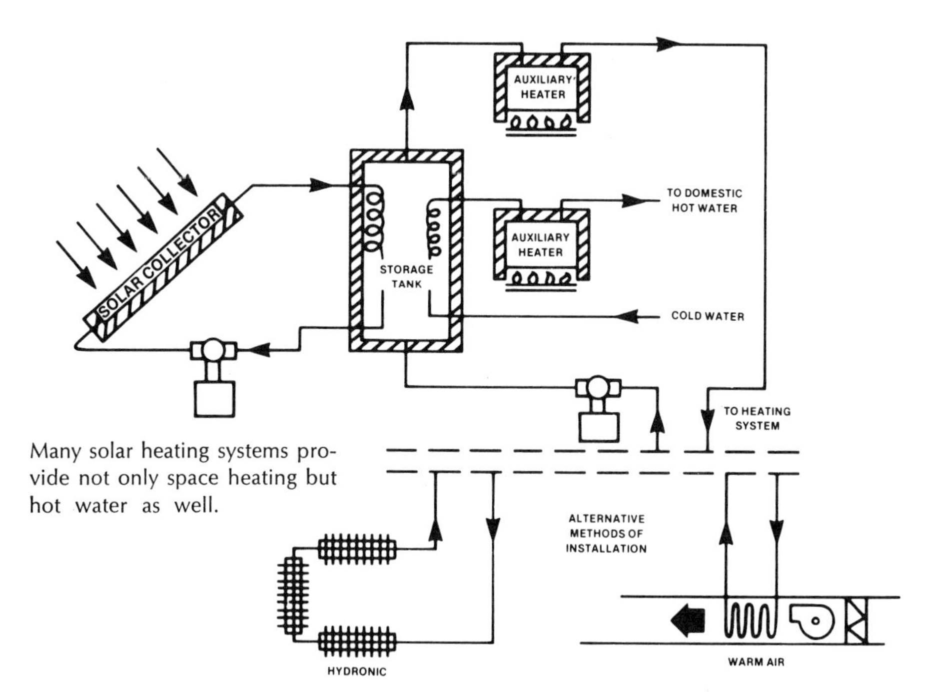

Many solar heating systems provide not only space heating but hot water as well.

REVERE COPPER AND BRASS INCORPORATED

Design. The optimum solar home must be designed with the sun in mind, with large windows on the south, and with small or nonexistent windows on the north.

The placement of the various components of the average household are somewhat different in a solar home. For example, the chimney may be placed internally rather than on an outside wall. This design innovation cuts down on potential heat loss. Entryways may be designed to protect the interior from quick influxes of cold air (or hot air in summer) from outside.

There are many different ways in which houses have been designed to make optimum use of solar energy for space heating. Many of these houses are still in the experimental stage, but others have been lived in for some time.

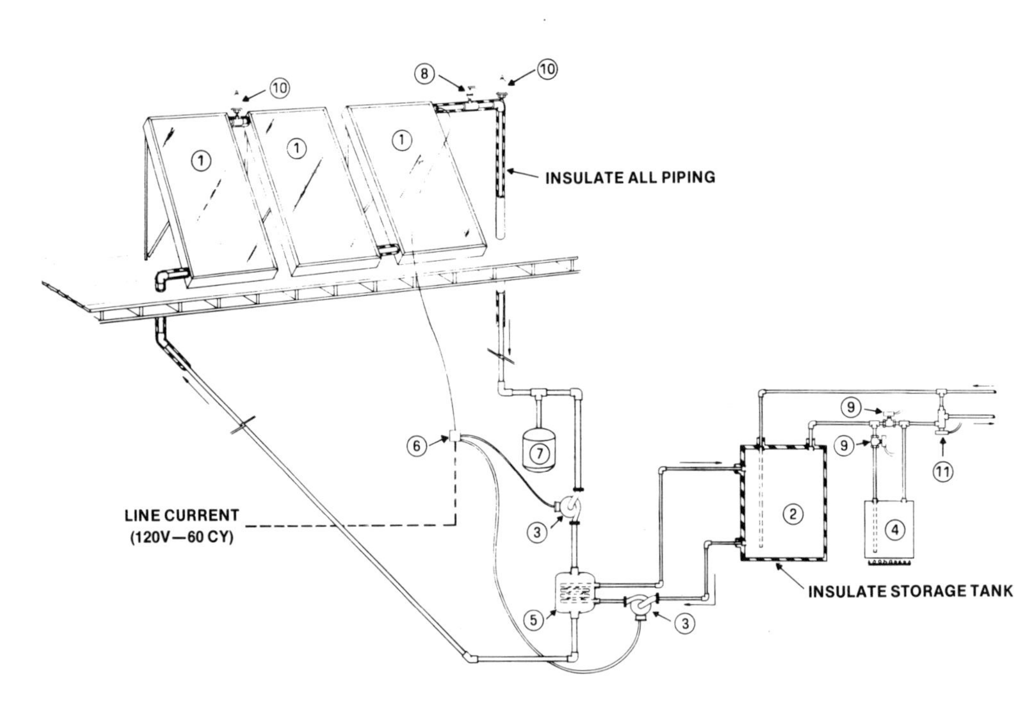

REVERE COPPER AND BRASS INCORPORATED

Solar apparatus includes these seven components. First, collectors (1) heat up medium which is moved to heat exchanger (5) by pump (3) at command of sensor (6). Storage tank (2) circulates water through heat exchanger (5). Auxiliary heater (4) provides backup warmth if necessary. (7) is expansion tank, (8) pressure relief, (9) shut-off valves, (10) air vent, and (11) mixing valve.

NASA

Solar collectors on top of steep-pitched roof of energy-saving experimental Tech House provide heat enough to cut fuel costs about two-thirds.

NASA'S SOLAR ENERGY TECH HOUSE

At Langley Research Center, Hampton, Virginia, the National Aeronautics and Space Administration is operating an experimental solar house called the Tech House. It is a contemporary-style home built to demonstrate how the average family can cut its fuel consumption by as much as two-thirds through the use of technological advances, including solar heat collection.

The house is of one-story design. It features a living room with fireplace, dining area, kitchen, two baths, laundry, two-car garage, and outdoor living room. Total enclosed living space is 1,600 square feet. Dr. Toss L. Gobel, chief of Langley's Research Facilities Engineering Division, and director of the experiment, believes that the house can be built within five years for about $40,000 to $50,000.

The use of solar collectors is one of the major features of the house, along with the use of night radiators and a heat pump. The combination of solar heat and the heat pump provides one of the most cost-effective heating and cooling systems now available.

The flat-plate, fluid-type collectors cover 320 square feet of the roof area, supplying some of the energy for space heating and domestic hot water. Excess solar heat is stored in a water storage system for use at night and on cloudy days.

Additional heat sources are controlled by venting large appliances like refrigerator, freezer, and wall oven inside in cold weather and outside in hot weather.

The Tech House is provided with plastic foam insulation installed as effectively as possible.

Heat pipe systems are used to recapture waste heat from the fireplace and return it to the house. Other advanced energy-saving systems include solid-state appliances.

The house faces east, but its solar collectors face south for maximum exposure to the sun. Extra areas of glass are located in the south windows.

Calculations show that the average contemporary house occupied in the area by a family of four uses the equivalent of 46,000 kilowatt hours of energy a year for central heating and air-conditioning, water heating, lights, kitchen appliances, television, furnace fan, and other equipment.

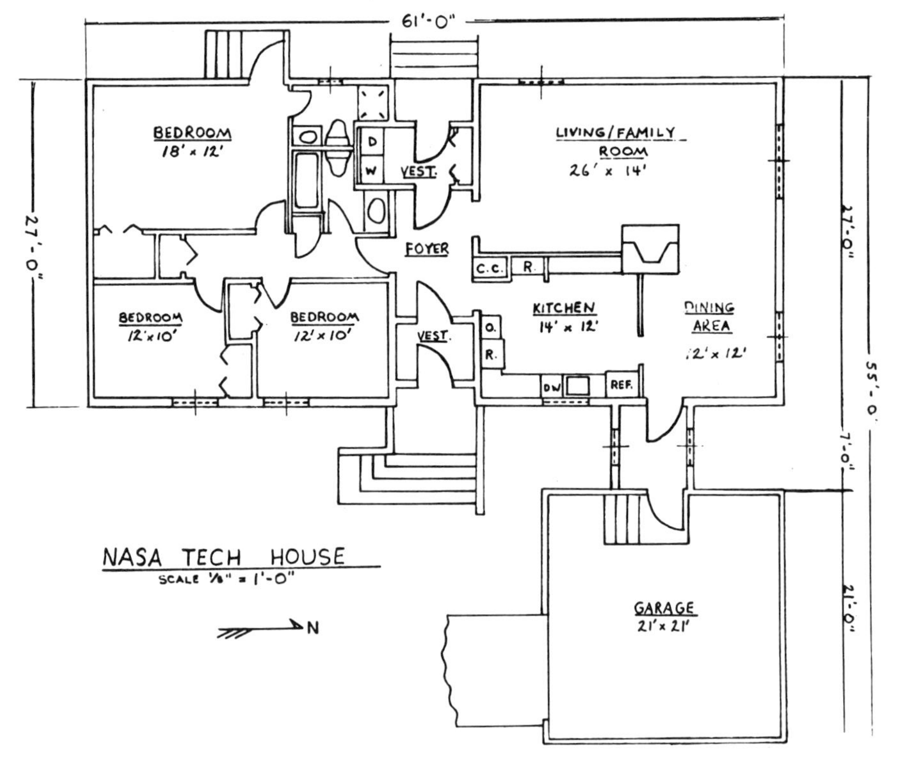

NASA

Floor plan of Tech House shows airlock vestibules at entrance and exit. Airlock port joins house and garage. Double doors prevent heat from escaping or entering.

Dr. Gobel estimates that the Tech House will use only about 15,000 kilowatt hours—less than one-third. The two-thirds saving in fuel bills will not only help conserve energy resources, but will pay for the initial investment of the house.

SPACE HEATING FROM A SOLAR WINDOW

A house in Bedford, New Hampshire, is operating as a completely passive solar house. No outside power is used to collect, store, or distribute the sun's heat inside the house. Natural convection and conduction currents move the heated air.

Designed from the start for solar heat, the house obtains 50 percent of its energy from solar collectors, 25 percent from sunlight passing directly in through the windows, and 25 percent from two wood-fired stoves.

The house is heavily insulated—even under the floor to isolate the interior from the ground below. All walls are likewise insulated. In addition, the north wall is covered with earth part way up from the ground to reduce heat loss from the north wind.

The house is built with foot-thick concrete walls to provide for as much heat retention as possible.

Windows are positioned to take full advantage of the sun. There are no windows on the north side of the house. The south wall is half insulated double-paned glass, floor to ceiling, and half solar collector.

Picture of New England house during construction shows combination of solar collectors (with vertical runnels) and conventional glass windows.

KALWALL CORPORATION

Each solar collector is a new type adapted especially for use in northern areas. Developed by the Kalwall Corporation, the collector has a typical rectangular configuration. It is composed of two fiberglass reinforced polymer face sheets bonded to both sides of a structural aluminum framework. The translucent sheets are spaced 2¾ inches apart, leaving a space of dead air between that acts as insulation.

Each sheet of Sun-Lite fiberglass transmits 88 to 90 percent of the light that strikes it. Both sheets together transmit 77 percent of the solar energy that strikes them.

Air space between the plastic sheets acts as a transmitter of light during the day. At night it becomes the repository of millions of styrofoam beads that are blown in between the plastic to serve as insulation at night or during overcast, cloudy, or stormy days, preventing the escape of heat through radiation. The addition of the plastic foam beads increases the insulation value of the panel fivefold.

When the sun is shining, the beads—each of which is about 1/16-inch in diameter—are pumped out of the space into a storage tank in the wall through pipes at the top and bottom of the panels. Then, when the sun sets or is behind the clouds, the beads are blown back into the panel. Control of the beads is automatic, although it can be accomplished manually.

The house is cooled in a natural manner by opening vents on the windowless north wall. Every door and window has insulating shutters that keep heat from escaping in or out. An airlock entry is provided from garage to house, composed of a pair of doors which keep the garage and house interior rigidly separated.

Domestic hot water is warmed by a preheater composed of a 75-foot length of polyvinyl chloride pipe embedded in the south "collector" wall. The sun's heat helps keep it warm.

For backup heat, the house uses two Austrian wood stoves. The three-bedroom house has about 2,000 square feet of living area.

A WALL OF WATER TANKS

One of the most startling and radical departures from the conventional type of solar house is that invented and built by Steve Baer in Albuquerque, New Mexico. Baer is the inventor, incidentally, of the type of solar collector described in the above New Hampshire house.

Baer's house is a variation of several other types of solar design. His solar collector, like the New Hampshire house's collection area, doubles as the home's south wall. It is not made of masonry or wood, but of 5-gallon oil drums filled with water!

The drums, stacked so the bottoms form the outer surface and the tops form the inner surface, are painted black on the bottoms and white on the tops.

Outside the water-filled drums is a plate-glass Trombe-like wall. During the day, the sun shines through the glass, strikes the black surface of the water drums, and heats the water. At night, clam-shell-shaped covers, called "skylids," are closed down over the drums to keep the heat from radiating out into the atmosphere. The covers contain louvers like Venetian blinds; they are lowered at night and raised in the daylight.

A SOLAR HOUSE ON LONG ISLAND SOUND

A research project that successfully provides 75 percent of a house's heat and domestic hot water requirements through solar energy is located in Quogue, Long Island, New York. The collector system, made by Owens-Illinois, has specially designed coated glass tubing which supplies hot water to a storage tank for computer-controlled distribution throughout the house.

The system, prepared for the newly built three-story 3,000 square foot home, is the design of solar and electronics engineer Joseph R. Frissora, an Owens-Illinois consultant.

Because of the large number of cloudy and overcast days on Long Island, the solar collectors used were selected because of their ability to absorb both direct as well as diffused sunlight.

The Sunpak collector is unaffected by wind and air temperatures, and can operate in areas of low sunlight from two to five times more effectively than most flat-plate solar collectors.

Triple-Tube Collectors Each 4-by-8-foot collector contains twenty-four glass tubes 2 inches in diameter and 40 inches long. In turn, each two-inch tube contains two more glass tubes, one inside the other. The larger inner tube acts as the absorber tube. It is selectively coated and hermetically sealed. The absorber traps the light energy from the sun and converts it to heat.

Air in the space between the outer tube and the absorber tube

OWENS-ILLINOIS, INC.
Solar system on this Long Island house collects 75 percent of space heating and domestic hot water requirements, then distributes heat by computer-controlled system.

Connecticut house on Long Island Sound gets about 65 percent of its heating needs from solar collectors on roof. Heating system is forced-air, with storage in 1,000-gallon basement tank.
SUNWORKS
(Photo: Robert Perron)

GRUMMAN CORPORATION
Vermont's Energy House supplies 50 percent of its total space heating and hot water needs through sixteen solar collectors mounted on sloping roof. Svstem works with antifreeze.

is exhausted and altered to a high vacuum, then permanently sealed. As such, it reduces heat loss, increases insulation, and prolongs the life of the collector.

Water heated in the Sunpak collector is transferred through a manifold to a 1,000-gallon fiberglass-reinforced plastic storage tank, especially designed by Frissora for Owens-Corning Fiberglas Corporation to meet the high heat requirements of the system.

The tank's composition of plastic makes it lightweight and inexpensive. Fiberglass will not corrode, preventing rust or other corrosive elements from entering the system.

Heat by Computer Control Frissora also designed the computerized control system that acts to distribute the heat from the storage tank to wherever it is needed in the household. The system provides for maximum use of the solar energy.

The household is divided into five heat control zones. A sensor in each zone relays temperature data to the computer, which automatically regulates the heat flow to that sector.

Sensors also monitor outside temperature, the amount of direct solar input, and the temperature of the storage tank.

When sunlight conditions are favorable, the system can store enough energy for a three-day supply of space heating and hot water.

If the house needs more heat than the solar collectors can provide, the computer control activates a system of valves and pumps that demand only as much heat as is needed from a backup tank heater operated by propane gas.

Heat is distributed throughout the house by means of a network of fans and coils. The entire house, or any one room, can be heated in from ten to thirty minutes, depending on the outside temperature.

Like all true solar houses, the structure is heavily insulated with 9½ inches of Owens-Corning Fiberglas building insulation in the ceilings and 6 inches in the walls. This is just about double the standard specification for residential installations.

The insulation prevents the escape of heat captured by the solar system. In turn, it adds to the general conservation of energy effected by the use of solar heat.

Because the house is a demonstration house, the computerized controls cost $100,000 to develop. Eventually, costs will drop to about $1,000 when enough of the units are in operation.

EXPERIMENTAL HOUSE IN MASSACHUSETTS

In Weston, Massachusetts, an inventive engineer has cut his potential fuel bills by more than 60 percent by using simple basic ideas in house design and construction. He doesn't have conventional solar collectors.

First of all, he has built his house, which he calls "Experimental Manor," in a normal way, with the only radical change a south-facing wall made entirely of glass. In addition to that surface, which serves as a solar collector, he has added other heat-absorbing materials inside the house, some fans to move the heat about, and has insulated the entire structure securely against heat loss.

Essentially, he has turned his house into a heat trap, with a south-facing window wall letting in the sun's energy, tight insulation storing it as heat, and fans moving it around the house to prevent the air from becoming stuffy.

Because he does not depend on expensive solar collectors and the construction of an enormous storage area inside the house, he is able to build a solar house a great deal less expensively than the kind envisioned by architects who use solar collectors, storage pits, and electronic controls.

In addition to the window wall, the house has concrete slab floors and walls built of pumice blocks. Both of these types of construction are excellent heat storers.

There is a slate mantelpiece, too, which absorbs and stores heat from the fireplace. Any stone will do the same. Wall studs, furniture, and plaster all become repositories of normal heat storage, acting in the same manner as the heat storage zone of a large water tank or a rock pit in a more complicated structure.

In Experimental Manor, room temperature rises when the sun shines in. When it gets 10° above normal, a fan turns on to blow air into the northerly rooms, using them essentially as backup storage rooms. When it becomes too hot in the south rooms, curtains cover the entire glass wall and seal out the sun's heat.

This home is a commonsense type of solar house. Unfortunately, the average person sometimes does not like to design a home with one whole wall completely of glass. And, sadly enough, it is not feasible to retrofit an ordinary house to the solar mode by tearing out the south wall and replacing it with glass.

SOLAR HEAT RUNS VERMONT'S ENERGY HOUSE

In New England, the homeowner spends about five years of his life just earning the money to heat his house. For the middle-income family, it is a problem of primary importance to cut down on fuel costs the year around.

For this reason, an experimental energy-saving middle-income residence was designed recently near Lake Quechee, Vermont. Its 2,300 square feet of living area makes use of as many energy-saving methods as are known to technology—without sacrificing spaciousness, adequate natural light, and the expected comforts and conveniences of modern life. And it satisfies esthetic and environmental standards as well.

And, most naturally, Quechee's Energy House is run by solar heat. Only winter heating is necessary. Summer temperatures of 65° need no considerations for air-conditioning.

Sixteen collectors manufactured by Grumman Corporation are used to capture the sun's energy, mounted on the sloping roof oriented 20° west of due south. The orientation was selected after tests showed that afternoon sunlight affords more heat than morning sunlight in the area. Cool, hazy mornings result in relatively poor collector performance compared to the warmer afternoon winter temperatures.

The latitude of Quechee is 43.5° North. Optimum winter solar collector performance is achieved by an inclination of 55°. However, the angle is too steep for maximum usage in the house, a pitch of 45° was agreed upon. The 10° difference does not interfere with energy collection.

Energy-Conservation Points at Quechee Structurally, the house features many energy-saving devices:

- All walls are insulated with 6 inches of thick insulation.
- The roof is insulated with 9 inches of thick insulation.
- The outer walls of the basement are insulated with 3 inches of insulation.
- All outside walls are sprayed with 1¾ inches of urethane, which acts as sheathing as well as insulation.
- The roof is covered with 1¾ overlapped urethane and then covered by asphalt shingles.
- All windows are triple-glazed with sealed gas sandwich for extra-strength insulation.
- The house is sited by a line of trees that provides a windbreak against winter wind.
- An airlock entryway prevents warm air from escaping during opening and closing of outside doors.
- A special solarium has insulated shutters to accept sunlight during the day and prevent its loss during the night.
- The house has no windows on the insulated masonry wall to the north.

In order not to sacrifice light in the north rooms, skylights are used to afford enough illumination so that no artificial lighting has to be used during the daylight hours.

The heating system is a combination heat pump and fluid-medium solar system. A forced-air ductwork network distributes heat to the rooms of the house.

The solar system supplies 50 percent of the heating and hot water energy needs of the house. The collector system will pay for itself in three to ten years, depending on whether it would be used in place of electric heating and hot water, or oil heating and hot water.

A SOLAR HOME YOU CAN HAVE CUSTOM-BUILT

Acorn Structures of Concord, Massachusetts, which has been in the business of supplying predesigned homes to buyers for years, has been experimenting with the design and construction of practical solar houses since 1974.

It now has come up with a practical solar home in which the heating system is integrated with the design and construction of the house. The house is marketed as a prefabricated home, with packages of precut material delivered to the site to be installed by professional builders.

The design has been tested out through one winter season. The 1,400-square-foot model will provide 46 percent of the space heating required during the cold season.

Solar Heating System The solar heating system is composed of four subsystems: one to collect solar heat, another to store it, a third to distribute it, and a fourth to provide backup heat.

The collector panels are made in large 4-by-20-foot plates. Each box is composed of a polyester-fiberglass cover, a black-painted aluminum absorber plate, and a copper-tubing grid underneath. The box has a 2-inch bottom of fiberglass insulation, a secondary waterproofing membrane, and ½-inch-thick plywood back that doubles as roof sheathing.

Heated water is stored in a 2,200-gallon water tank lined with vinyl and framed with wood. Energy enough to warm the house for a few days can be stored in the tank, but it takes several days of bright sunshine to charge it.

Heat is delivered to the rooms of the house by pumping warm stored water through an open coil placed in the duct of a warm-air heating system. A blower moves air across the coil and through the ductwork. If the coil is not hot enough, the backup heater provides heat for the ductwork. The backup furnace is a conventional oil-fired warm-air furnace.

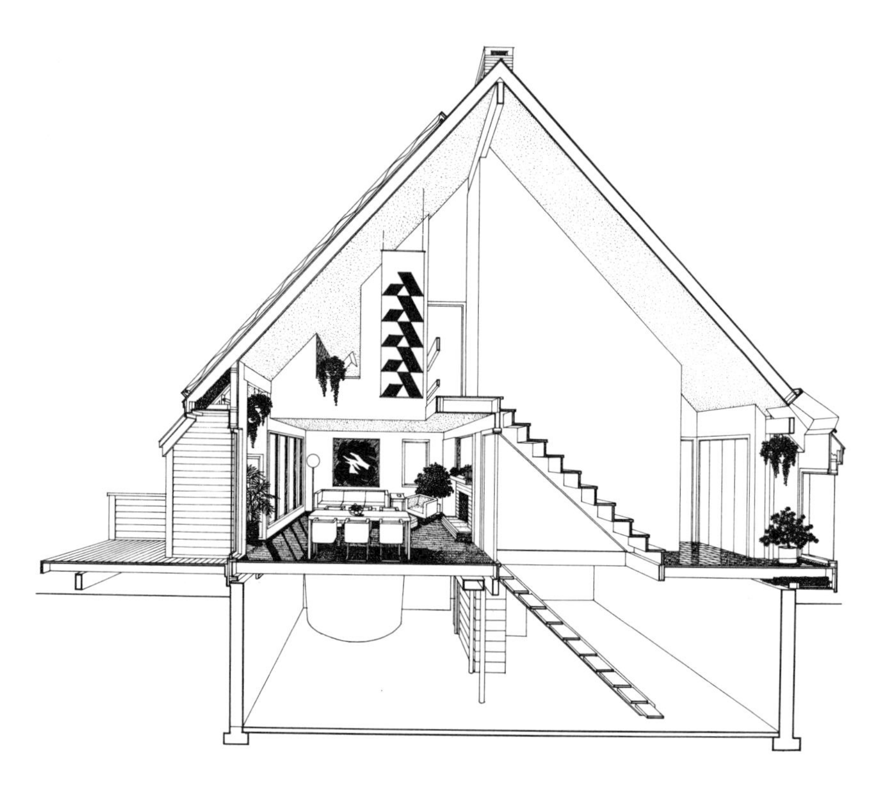

ACORN STRUCTURES, INC.

Artist's rendering depicts solar house by Acorn Structures. Sun provides 46 percent of space heating requirements and most of hot water needs. Accompanying diagram shows side view of interior.

Heating is controlled by a thermostat. When the house cools to the temperature at which the thermostat is set, a pump is activated and heated water is sent from the storage tank through the open coil. Air blown over the heated coil is warmed and is distributed through the house until the thermostat is satisfied and goes off.

If the solar-warmed water is not hot enough to satisfy the thermostat, the house will cool until it activates the second-stage

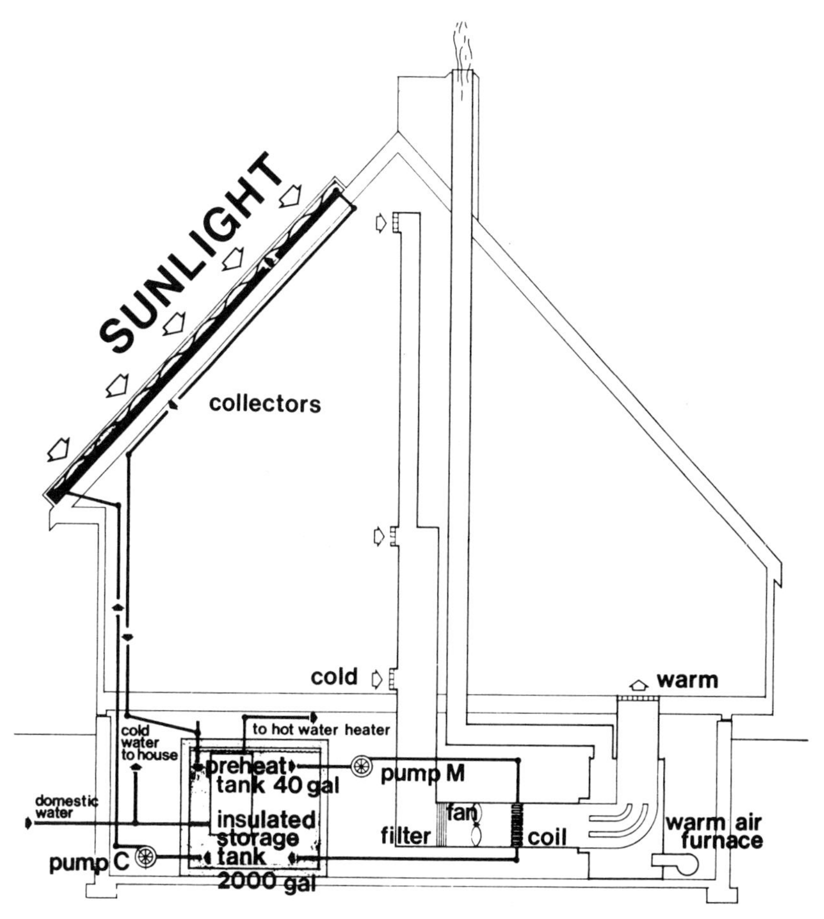

ACORN STRUCTURES, INC.

Collectors bring solar heat down from roof and impart it to storage tank through heat exchanger. Air blown across hot coil carries warmth up into house through duct.

sensor, which in turn will fire up the backup furnace to heat the house to its proper temperature.

In cloudy winter weather when the stored solar energy is being rapidly depleted, blower and pump will run for long periods. However, a secondary effect of the blower is to distribute immediate passive solar heat that comes in through southern-exposure windows. This heat keeps the stored energy from running out too quickly. A third effect of the blower is to move the new heat from the south side of the house to the cooler areas on the northern side.

The Collector A sensor measures the temperature of the water flowing through the solar collectors. When the fluid is determined to be sufficiently warm to add its heat to that of the stored water, a pump turns on, drawing the water down to the storage tank. When the pump is not working—for example, at night when the sun is not out—the water simply drains out of the collector back to the storage tank, protecting the collector from freeze-up.

The house's supply of domestic hot water comes directly from a preheated 40-gallon tank. The preheated tank is supplied by water from the storage tank. The solar collectors usually supply enough heat to raise the water tank temperature over 120° and to carry almost all the domestic hot water load.

In the active space-heating months of winter, the temperature of the storage tank may be only in the 70° to 80° F. range. Water coming in to the system is much colder than that. Preheating the fresh water to even those temperatures will represent a very significant gain in the domestic hot water temperature and will play a significant part in the hot water heating supply. Backup heating comes from an electric heater.

Solar House Design The entire Acorn house is designed with solar heating in mind, particularly in regard to the amount of solar heat allowed in through its windows.

It is also designed and constructed to cut down heat loss through the walls, through the roof, through the floors, and even by way of infiltration through open doors.

That means that the house has maximum insulation in walls, floor, and ceiling. Double-insulated glass is used throughout the house, with windows and doors tightly weatherstripped.

The design of the house specifies small windows on northern exposures, large windows on southern exposures, and large protec-

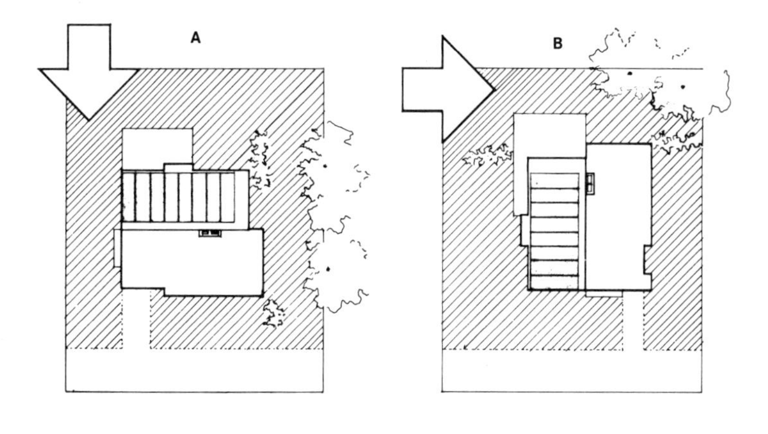

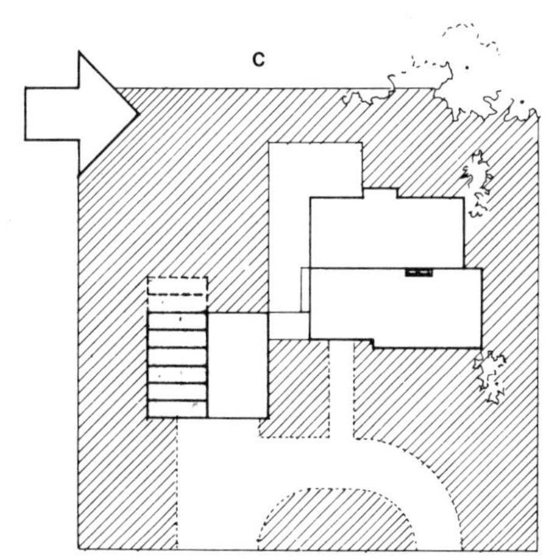

Diagrams show how house, though oriented to south, must respond to natural conditions on site. In (A), house is placed with tree to west, east, or north. In (B), trees to west and northwest and at windward corners help reduce heat losses to prevailing winds. In (C), collectors are mounted on garage, or can be set up on any other detached structure.

ACORN STRUCTURES, INC.

tive overhangs. Living areas are located so that they benefit directly from the southern exposure.

Bedrooms and hallways are placed on the northerly side of the house or on the second floor, where they can be seperately zoned and regulated at lower temperatures. Buffer zones are provided for areas that need to be maintained at higher temperatures.

The pitch of the roof surface on which solar collectors are mounted is 47°. Why 47°? A 35° roof pitch is not steep enough to allow snow to slide off quickly, and a pitch steeper than 47° makes for unusable space under the peak.

The Acorn house is designed for siting northerly climes, of course, but can be used anywhere. The New England area has fewer bright days than the Southwest, and consequently cannot deliver as high a percentage of solar energy.

OTHER SOLAR SPACE-HEATED HOUSES

• In Stow, Massachusetts, one solar energy house that gets most of its heat through solar collectors keeps the temperature at a comfortable 68° to 70° F. even during the coldest days of the New England winter.

The sun's heat supplies the house's space heating needs and its domestic hot water heating needs, cutting fuel bills by about 75 percent.

The house has an attached swimming pool, which is also heated by energy from the sun.

The system involves fluid-medium collection of the sun's heat, in which the fluid is actually a mixture of water and antifreeze. Heated by the sun to around 200°, the fluid circulates down in a connecting system into a heat coil inside a 2,500-gallon storage tank. The solar heat is imparted to the storage tank.

When warm air is needed in the house, hot water circulates through a heat coil near an air duct, where air is blown over the coil and then ducted to every room in the house.

• A solar house in East Hampton, New York, has a unique solar storage system located in the roof. Conventional solar panels first collect heat which is pumped into a roof space containing 1,000 sealed plastic bottles of water. The water holds heat enough to provide for space heating for three sunless days.

The system supplies between 50 and 75 percent of the home's heating requirements.

• In a year-round beach house in Connecticut, a combined active and passive solar heating system provides about 65 percent of the house's total heating needs.

The roof contains enough solar collectors to heat 2,000 square feet of living area in the house.

Storage is accomplished in the basement, where a 1,000-gallon tank converts the solar energy to heated water, and then provides warmed water in a coil over which filtered air is blown to carry heat to all rooms through the house's ductwork.

In spite of fogs and cloudy weather, the system provides about 65 percent of each year's heating needs.

8

Solar Air-Conditioning

The cooling of a home by means of solar energy is as technically practicable as heating it. Nevertheless, the techniques of air-conditioning by solar energy are not so advanced nor in such widespread use as those of heating.

Several drawbacks are immediately apparent in planning the cooling of a home by the sun. At a rough estimate, the area of solar collection required to cool is just about twice that required to heat.

While the collection area for space heating is equal to about half the living area, the collection area for cooling is equal to the living area.

In a flat-roofed, one-story house, almost the entire roof must be covered by collectors. However, the cost of nonsolar air-conditioning, most of which is accomplished by electric power, is considerable. The cost of solar air-conditioning, once the equipment is installed, is extremely low.

PREPARING A HOUSE FOR SOLAR COOLING

The architecture and design of a house can make a big difference in its ability to remain cool in the summertime.

Windows or vents on the north side of the house can help. Sloping roofs with long overhangs can keep the windows in shadow during the summer.

Light paint which reflects sunlight and heat can cause a great deal of relief from heat that might otherwise be absorbed in the walls.

And, generally speaking, insulation is just as important for keeping a house cool in the hot weather as it is for keeping it warm in cold weather.

Recent studies by heating and cooling engineers show regional differences in the relationship between heating and cooling efficiency in solar systems.

In a coastal region with a great deal of warm weather like Los Angeles, California, solar heating can provide more than 100 percent of a home's heating needs throughout the year. It can also deliver 98 percent of a home's cooling needs. (It usually takes about twice as much energy to cool a house as it does to heat it. In a warmer region like the Southwest, heating in the winter is really not much of a problem.)

In Atlanta, Georgia, solar energy can provide about 85 percent of a home's needs for heating, but only about 63 percent of its cooling needs. (The winters are mild and the summers hot.)

In New York City, New York, solar energy can provide about 84 percent of a home's heating needs and about 77 percent of its cooling needs. (New York is cold in the winter and sometimes quite hot in the summer.)

In Seattle, Washington, however, solar energy can provide 100 percent of a home's *cooling* needs but only 46 percent of its *heating* needs. (Obviously Seattle has fairly mild summers and cooler, cloudier winters.)

In Chicago, Illinois, solar energy can supply 76 percent of a home's cooling needs, but only 50 percent of its heating needs. (Chicago does not get terribly hot in the summer, but is cold and overcast in the winter.)

In Miami, Florida, solar energy can provide 92 percent of a home's cooling needs, but only 45 percent of its heating needs. (The summers are warm, but there is a lot of sunshine; in the winter, the weather is fairly mild, but there are not that many bright days.)

The subject of regional differences in the relationship between heating and cooling efficiencies of solar systems is explored at further length in Chapter Fourteen, with more specific figures

ALUMINUM ASSOCIATION
All-solar energy development in El Cajon, California, is heated and cooled by sun. Spanish-type tile roofs have flat areas at rear where collectors are installed.

regarding "insolation," the amount of sunshine available from month to month.

In spite of the discrepancies as to efficiency in various regions of the country, solar energy can be used effectively to cool homes, although most systems must provide backup machinery for current installations.

Experiments are in the works on advanced types of cooling techniques which may someday be adapted for use in conjunction with solar heating systems. So far, there is only one type of cooling system commercially available for direct use with solar collection systems. That type is the absorption cooling system. It is the only kind of cooling system that utilizes heat energy directly.

HOW AN ABSORPTION AIR-CONDITIONER WORKS

Although Freon gas is the most familiar refrigerant used in commercial freezing units (see Chapter Four), water can be used for cooling too. Freon has an extremely low boiling point, causing water to turn into ice. Water has a higher boiling point. For use in cooling above-freezing temperatures, water is more efficient.

It is more efficient as a refrigerant at high temperatures because of the greater amount of heat it withdraws from its surroundings at vaporization.

When ammonia, for example, condenses, it gives off 327 calories. Water gives off 537, almost 40 percent more.

The vaporization of water can be used effectively for the cooling of air in the home. Because vaporization can be achieved easily by the use of solar energy to raise the heat of liquid, the most effective type of solar energy air-cooling system uses water.

The typical absorption system utilizes a combination of lithium bromide and water to produce cooling. The lithium bromide solution, diluted by the absorption of water vapor, is regenerated by heat to give off a highly concentrated solution.

It works this way. Lithium bromide has a strong affinity for water. That means that it absorbs it readily. A generator is filled with a solution of lithium bromide acting as the *absorber* and water acting as the *refrigerant*.

Solar-heated hot water direct from the roof collector flows through a heating coil at a temperature of 200° F. The lithium bromide outside the coil vaporizes and boils violently. This mixture of lithium bromide spray and droplets of water is shot up through a vertical tube.

Gravity causes the lithium bromide to drop off and flow down. The water vapor, which is lighter, rises to a condenser coil, and becomes separated from the lithium bromide by means of a series of baffles.

The water vapor in the condenser coil passes through a tank of water piped in from a cooling tower, usually located on the roof. The cool water condenses the water vapor to liquid form once again.

In the condenser, the air pressure is about 1/14 of an atmosphere—50 to 60 millimeters of mercury, compared to the normal atmospheric pressure of 760 millimeters.

The water now passes through a stricture into a cooling coil where the atmospheric pressure is reduced dramatically to 6 to 9 millimeters of mercury. This tremendous drop in atmospheric pressure reduces the boiling point of water, causing it to boil at a low 40° F., producing refrigeration as the process of vaporization quickly absorbs heat from the surrounding area.

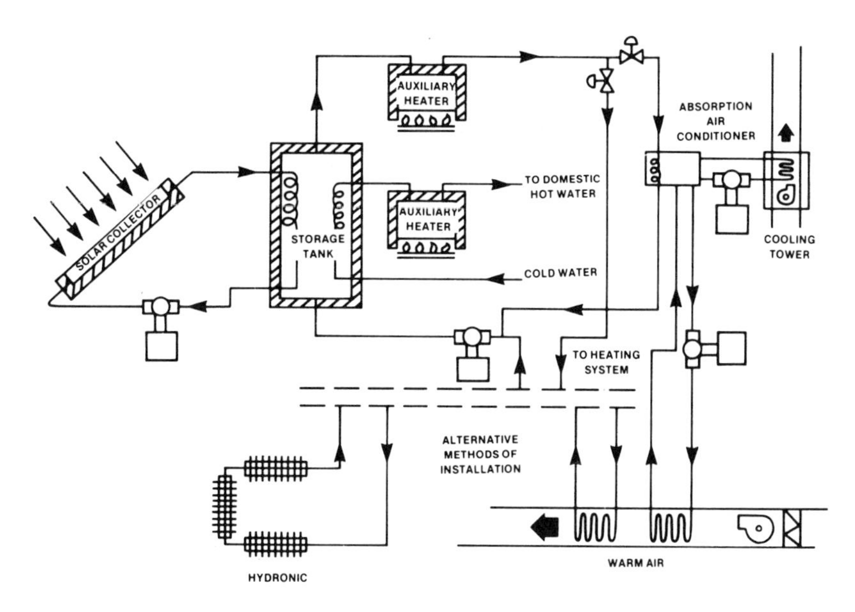

REVERE COPPER AND BRASS INCORPORATED

Typical installation for solar heating, cooling, and hot water includes absorption air-conditioner tied in with collector system. Note cooling tower, part of absorption unit.

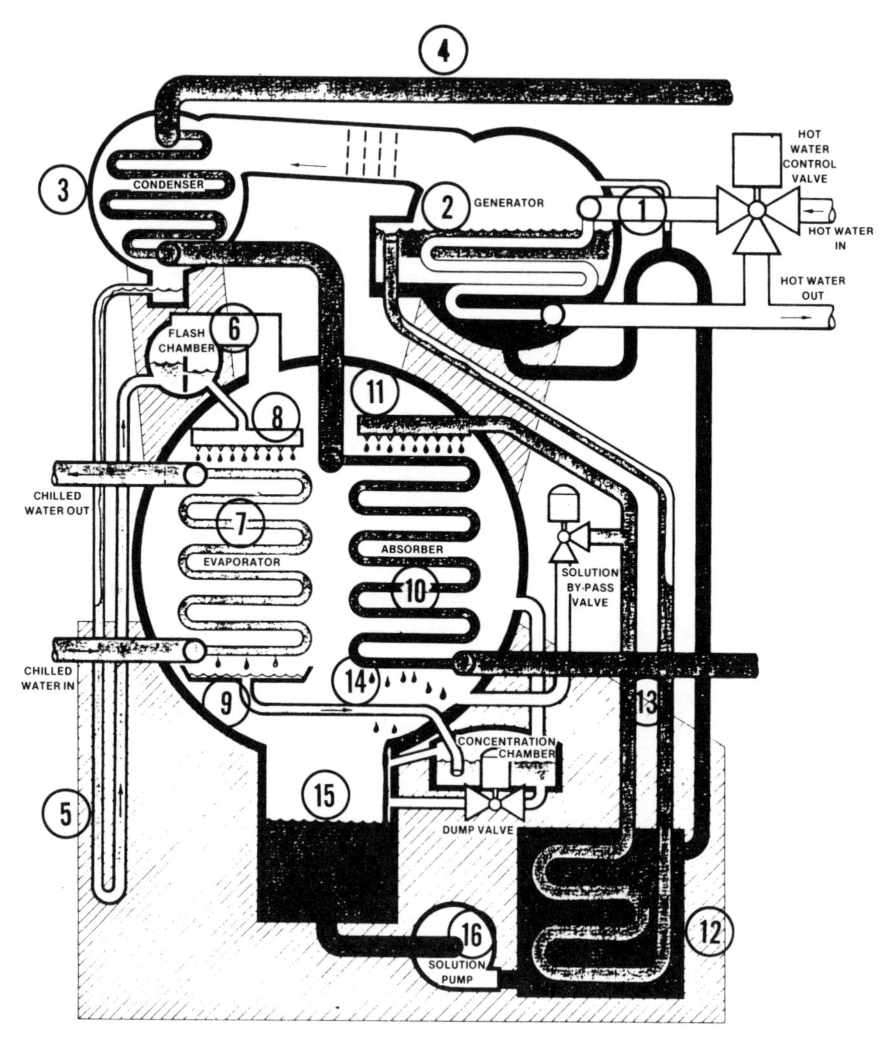

ARKLA INDUSTRIES

Solar heated water enters generator tubes (1). Heat vaporizes part of water refrigerant in generator (2), separating it from lithium bromide absorbent. Vaporized refrigerant passes to condenser (3), giving up latent heat to condensing water (4), and is liquified. Refrigerant flows through U tube (5) to flash chamber (6) and to evaporator (7), wetting outside surfaces of evaporator coil tubes (8). It is again vaporized, absorbing heat of refrigeration load from chilled water (9). Vapor flows to absorber (10) where it is reliquified, combining with absorbent (11). Hot absorbent goes from generator (2) to liquid heat exchanger (12) where it loses heat. Pre-cooled absorbent enters absorber (13), wetting outer surfaces of absorber coil tubes, combining with vapor refrigerant. It gives up remainder of heat to condensing water flowing inside absorber coil tubes (14). Then it flows to solution pump (16) which moves it through liquid heat exchanger (12) where it is pre-heated to continue on to generator (2) and repeat cycle.

Air in the vicinity is cooled for delivery to the areas of demand.

Once again the water vapor proceeds to the absorber, where the concentrated lithium bromide solution reabsorbs it. After taking on the water vapor, the lithium bromide flows through a heat exchanger, where it is partially heated and is returned once again to the generator to go through another cycle.

This mixture of lithium bromide and water can be used in continuously operating air-conditioners. The amount of heat captured by solar collectors is sufficient to keep the mixture working easily. Regeneration of the solution is accomplished with water warmed only to 170° F. The temperature is below the boiling point of water. It is low enough to be supplied by the solar collection of a conventional flat-plate solar panel.

In a test conducted by Arkla Industries of Evansville, Indiana, a manufacturer of a lithium bromide-type absorber air-conditioner, an air-conditioning unit was connected to a solar collection area.

The collectors totaled 102 square feet of solar area. Rooms covering 1,200 square feet in area were cooled for several days with the unit.

About 30 percent of the solar energy collected was transmitted to the circulating water solution during the period of operation.

Another test measured energy collected by 200 square feet of collection area. On a bright, calm day, an air-conditioning unit produced 1 ton of refrigeration (12,000 BTUs per hour).

SOLAR HEATING/COOLING TEST HOUSE

An Arkla air-conditioning/heater unit is installed in the Decade 80 House in Tucson, Arizona, a solar energy house designed and constructed by the Copper Development Association as a test for solar heating and cooling.

The heating/cooling system is designed to supply solar energy by means of fluid-medium solar collectors operating at temperatures between 190° and 210°.

Most of the solar energy collected is used in running the house's cooling system. However, the system is designed to supply space heating, domestic hot water heating, and swimming pool heating in addition.

CALIFORNIA REDWOOD ASSOCIATION
Copper fluid-medium solar collectors provide 75 percent of this desert home's cooling and 100 percent of its heating.

DECADE 80 HOUSE SOLAR ENERGY SYSTEM

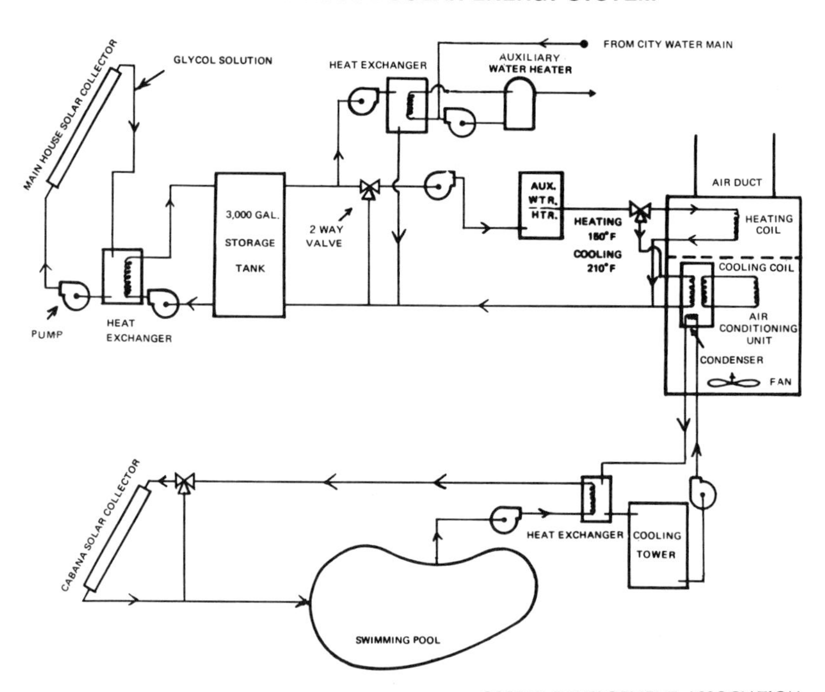

COPPER DEVELOPMENT ASSOCIATION
Schematic plan shows how solar system can supply energy for the use of absorber-type air-conditioning system sited in hot climate.

Air-conditioning is provided by two standard Arkla lithium bromide water-absorption units modified for hot water firing. The two units together have a total cooling capacity of 72,000 BTUs per hour. They are also capable of supplying space heat during the winter and heat for hot water the year around.

A one-story, three-bedroom dwelling with a swimming pool attached and a small guest wing nearby, the Decade 80 House faces the south, with solar panels integrated in its south-sloping roof. Separate solar panels, mounted on the guest-wing roof, supply heat for the swimming pool.

The guest-wing roof slopes 40°, favoring pool heating during spring and fall when the sun does not warm the water sufficiently. The house roof slopes 27° to favor energy collection for summer space cooling needs.

The Collector Subsystem The collectors, which cover an area of 1,800 square feet, are copper flat-plate panels, each made up of two sheets of glass, a copper solar absorber plate painted black, rectangular copper tubes through which the fluid medium flows, and back insuluation of reflecting foil and 3½ inches of fiberglass.

They are approximately 2 feet wide and run from the eaves to the ridge of the house. All are bonded to plywood ⅜-inch thick.

The copper tubes and copper fins in the collector are blackened to convert the sun's light to heat. The fluid medium in the conduit is composed of a propylene glycol–water solution, which flows through a closed circuit from panel to storage tank and back.

Solar heat is transferred from the circulating medium to the water in the tank by means of a heat exchanger. Heat is drawn from the tank as needed for domestic hot water, for space heating, and for space cooling.

Swimming pool panels are identical to house panels, except that they are not covered with glass.

The Storage Subsystem The average amount of storage requirements for an effective solar cooling system is somewhere between 5 to 10 pounds of water for each square foot of collector area. That translates to from 9,000 to 18,000 pounds for the 1,800 square feet of collector zone. In this conservative setup, 2,800 gallons is used for optimum storage, which translates to 22,400 pounds—4,400 over the optimum.

Water is stored in a 3,000-gallon cylindrical tank approximately

8.5 feet in diameter and 8.5 feet in height. It can deliver 450,000 BTUs for the cooling system, and up to 3,330,000 BTUs for the heating system.

The air-conditioning unit will operate at full capacity for about four and a half hours. Almost thirty hours' worth of cooling supply is available with the solar storage tank at full capacity.

The heating/cooling system will operate at capacity with collector temperatures down to a little less than 190° F.

Control of Heating/Cooling System Three major control systems operate the house: the solar collector and heat storage subsystem; the hot water distribution subsystem; and the swimming pool heating subsystem.

Conventional thermostat controls operate the entire solar collection and distribution system automatically. For summer cooling and for winter heating, however, manual selection is needed to switch from one mode to the other, depending on outside weather conditions.

Test results from the first year of operation of the Decade 80 House show that the system provides essentially 100 percent of the solar power needed to heat the house and a little over 70 percent of the power needed to cool it in the summer.

9

A Heating/Cooling System

Because the absorption air-conditioner is not yet efficient enough to carry 100 percent of the cooling load of a solar home, engineers have been experimenting with other means of space cooling.

A PASSIVE HEATING/COOLING SYSTEM

There are several truly passive systems of solar heating and cooling now in use. Some of them do not use conventional solar collectors at all, but employ other means of capturing heat and moving it from where it is not wanted to where it is wanted.

One of the most successful and commercially feasible of these is Harold R. Hay's Skytherm house design. Hay began developing his system back in 1954 in New Delhi, India, where he was trying to warm and cool houses without use of electricity or other types of costly heating and cooling fuels.

His system is basically very simple. It involves only a solar collector and a control system to distribute the heat.

A Pond on a Rooftop The collector is simply a pond of water built on top of a flat-roofed house. The control system is a movable insulator that can be operated over the pond to cut down on the absorption of solar energy in the daytime and to regulate the evaporation of water in the nighttime.

This method of heating and cooling is designed for a house in a hot climate which during the average year demands cooling more than heating.

Design of the house is a primary consideration. One of Hay's prototypes is a simple flat-roofed house with carport area attached in Tempe, Arizona.

The test house contains a room 10 feet by 12 feet. The water pond on the roof is actually a series of 2-foot-square containers, extending 12 feet in one direction and 15 feet 5 inches in the other. The longer section extends out over an attached carport. The basins contain about 7 inches of water. Part of the pond is covered, and the rest can be open or closed on demand. Insulation panels move on rollers to the carport section when the pond is open.

The walls of the house are constructed of earth and brick in the manner of a desert house, thick enough to provide good insulation and regulation of temperature. The floor is a 4-inch-thick concrete slab. Walls and floor store enough heat from the daytime to warm the house partially at night, even without the heat from the roof collector.

Collector ponds are supported by corrugated steel laid on the ceiling structure, covered on top with 2 inches of polystyrene insulation and lined with black polyethylene sheets 10 mils thick.

Three insulation slabs, each 4 feet by 8 feet and an inch and a half thick, are made of rigid urethane. Their total weight is 125 pounds. The slabs are moved on nylon rollers, operated by hand or mechanically.

Moving the Insulator Panels To start out an average winter day, the insulation slabs are removed from the ponds in the early morning when the sun first comes up. The solar rays pass through the transparent water in the pond, commencing to heat the black plastic at the bottom. This heat moves into the house, warming the room air at the top.

Actually, the temperature of the pond's bottom heats up faster than the water does. As the day's heat increases, the pond stores up heat for the night. Temperature is evenly maintained in the house by using the heat in the pond on demand.

At night in cool weather, the insulation panels cover the ponds to prevent the heat from radiating out into the atmosphere.

In summer, the insulation panels cover the ponds in the day to keep out the heat, and are pulled back at night. Evaporation at night cools the water, in turn cooling the house. In the daytime, a blower hastens evaporation, supplying cooling for the house.

In effect, the low-efficiency collector for winter heat becomes a high-efficiency heat radiator for summer cooling.

The amount of heat and cooling is controlled by the movement of the insulation back and forth over the ponds.

A temperature of 70° to 80° is maintained each day, 90 percent of the time. The rest of the time, the temperatures fluctuate only from 68° to 82°.

The only two additional controls used in the experimental house are a blower that sweeps air across the water during the three summer months to hasten evaporation and cooling, and a fan that sends this air into the house.

As can be seen, the house is warmed and cooled by solar energy trapped in the roof ponds. The amount of heat used from the ponds rises and falls to accommodate heat gain or loss from the room. A movable external insulation panel is used as a valve maintaining the heat stored in the ceiling ponds.

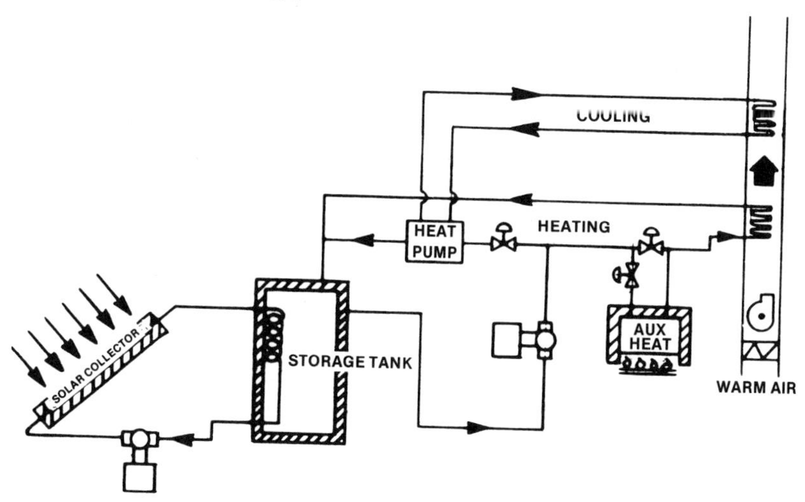

REVERE COPPER AND BRASS INCORPORATED

Heating and cooling system is supplied by solar energy in conjunction with heat pump. Pump uses solar warmth to move heat in during winter and out during summer.

Motor to move insulation panels runs three minutes morning and night to move panels across Skytherm's rooftop pond of water.

SKYTHERM PROCESS AND ENGINEERING

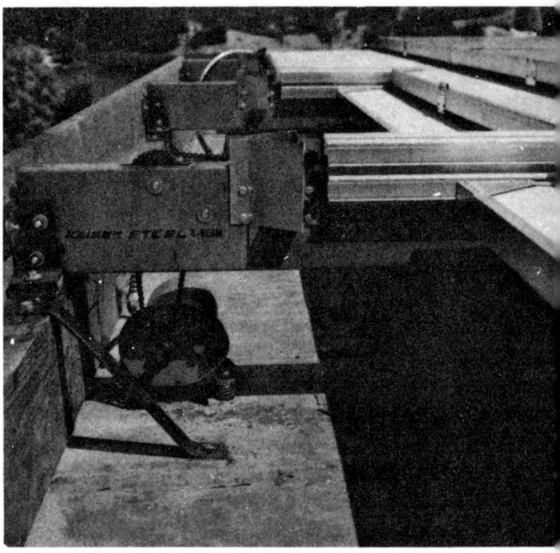

In addition, heat in the 4-inch-thick concrete slab floor substantially supplements that in the ceiling ponds, and simply rises into the room from the floor surface.

HEATING AND COOLING WITHOUT COLLECTORS

Several different types of solar heating and cooling systems have been devised which do not use commercial solar collectors.

One of them is a home in Albuquerque, New Mexico. The builder wanted to make use of indigenous materials in the building of the house: he chose adobe and logs.

Built of adobe bricks, the residence has walls constructed 14 inches thick. A total of 150 tons of adobe is used in the dwelling. Adobe, like most masonry, is a fairly bad insulator, but an excellent heat-retention material.

Structural support for the roof is afforded by rough-hewn logs.

Homemade solar collectors form part of the pitched roof of the house. They are made out of black-painted sheet metal nailed to the rafters of the house. A glass cover retains the heat in the collector box. Metal lath placed between the absorber plate and the cover breaks up the air flow and provides more metal-to-air heat transfer.

The roof structure beneath the collectors is tightly insulated.

For solar storage space, a pile of rocks is buried below the ground floor, which is constructed of laid brick. Warmed air from the collectors is blown by ducts down to the rock pile. The storage zone is located under the north side of the house, beneath the living room.

Heat is provided by two methods: by radiant heat through the brick floor; and by means of a fan-controlled air duct connected with the rock pit.

The south windows of the house are protected from direct sunlight in the summer by a trellis supporting climbing plants with large leaves and petals.

In the winter, the plants lose their foliage, letting in direct solar heat through the double-paned glass windows.

The combination of adobe wall construction and roof-mounted solar collectors provides enough energy to heat the house and store heat for five days at 0° F. outside temperature.

Generally speaking, the leaves of the plants provide the maximum protection against heat in the home. The double-panel insulated glass windows let in a certain amount of heat, but the adobe walls, being 14 inches thick, store up the day's heat and then radiate it into the darkness when the atmosphere cools after dark. Walls of that thickness, once commonplace in the Southwest, usually keep the inside temperatures of the house at least 10° F. below that of the outdoors. They are a natural method of controlling the summer's heat.

SOLAR ENERGY HOUSE THAT WORKS ON SENSOR CONTROL

In Mission Viejo, California, an experimental house utilizes solar energy in part and cuts down on heat escape in part.

A joint venture of the Energy Research and Development Administration, the Southern California Gas Company, and the Mission Viejo Company, the new home is called the Minimum Energy Design House.

The house design markets at about $45,000. It is a three-bedroom house, with living room, country-style kitchen, bathroom, and garage.

Main feature of the house is its maximum insulation properties. But its design makes use of solar energy as well, although not with conventional solar collectors.

The tile roof overhead—made of terra cotta curved tiles in the

Spanish tradition—helps hold in the heat by storing it on sunshiny days. Walls are of masonry, covered with stucco surfacing. The stucco, painted white, reflects the sun's heat in the bright hot days, keeping the interior cool.

Buried inside each wall is a plastic membrane that cuts down on the amount of air that can filter through, either on its way in or on its way out. Each wall is 6 inches thick and crammed tight with insulation.

To keep the house cool in the summertime, the roof extends out further than normal in a long overhang to shade both windows and walls. Windows are of double-paned glass. Sandwiched in between the double panes of glass are small louvers like Venetian blinds that never have to be dusted. They can be opened or shut to control the entrance of heat in the summer and the escape of heat in the winter.

Inside the house several innovative energy-saving ideas have been incorporated. For example, the refrigerator vents its waste heat outside the house in the summer. In the winter, the waste heat is vented into the heating system to help keep the house warm.

A special hot water system allows the operator to select the exact water temperature he wants, rather than waste a lot of hot water while adjusting for the desired temperature. A calibrated hot water tank serves the system.

As in the New Mexico adobe house, the thick walls are a throwback to an earlier and simpler system of building. Not only do these walls keep the house 10° cooler on the hottest California day, but they also cut off outside noises—even the racket made by a tractor on the front lawn.

The heating and cooling system is controlled by a complex and innovative system developed by Honeywell Corporation. The elaborate control box is located at one end of the garage.

From it rises a confusion of pipes, tanks, and ducts. Sensors inside and outside the house read temperatures constantly.

If the outside air is warm and the inside air is cooler than the thermostat calls for, the blower moves outside air into the house to heat it.

If the outside temperature drops in the summertime, and the interior of the house stays too warm, the blower moves the outside air inside to cool the house.

The house will cut down by at least 70 percent on the consumption of fossil fuel.

HEATING AND COOLING WITH A HEAT PUMP

In Coral Springs, Florida, Electra III, an experimental house operated by Westinghouse Electric Corporation, combines a heat pump air-conditioning system with an active solar heating system to take care not only of space heating, space cooling, and domestic hot water, but swimming pool heating as well.

The heat pump is the key to the system. Although it operates on electric power, it utilizes much of the solar energy gathered by the collector panels on the roof to heat and cool the house, to heat the domestic hot water, and to warm the pool.

Because of its location, the house is designed and oriented primarily to minimize the need for air-conditioning. Structural features in the design of the house cut down on the need for cooling.

While the roofs of most Florida homes have 2 feet or less overhang, the Electra III house has a 4-foot overhang all around. The patio area has 20 feet of coverage over it. This keeps sunlight off the

WESTINGHOUSE ELECTRIC CORPORATION
Florida house uses ingenious combination of solar energy and heat pump to heat, cool, supply domestic hot water, and warm its swimming pool.

WESTINGHOUSE ELECTRIC CORPORATION

Four-foot overhangs effectively block out sunlight during summer months when sun is directly overhead.

windows and the walls during hot weather when the sun is high in the sky. Solar heat on walls and windows is quite often the main reason for the need of air-conditioning.

The house has high roof hips, capped with ventilating outlets. The use of these ventilators increases natural attic ventilation ten times more than louvered vents, and reduces attic temperatures considerably, lessening the need for cooling in the house below.

The house has 10-foot ceilings throughout, except in the living room, dining room, and kitchen areas, where they are 12½ feet high. High ceilings allow unwanted heat to rise above the space occupied by people.

To reduce the number of days that air-conditioning is needed, the house is oriented to take advantage of the southeasterly breezes that prevail during spring, summer, and fall.

Breeze sweeps across the swimming pool and heavily shaded patio before entering doors which open to the south and east. Moving into the house, the breeze removes the heat which has gathered at the ceilings and carries it out the windows and doors to the west and north.

All walls and ceilings are insulated as efficiently as possible to reduce the need for cooling during hot weather and for heating during cold weather.

Windows and glass doors are bronze-tinted and double-insulated, set into carefully sealed frames. This construction tech-

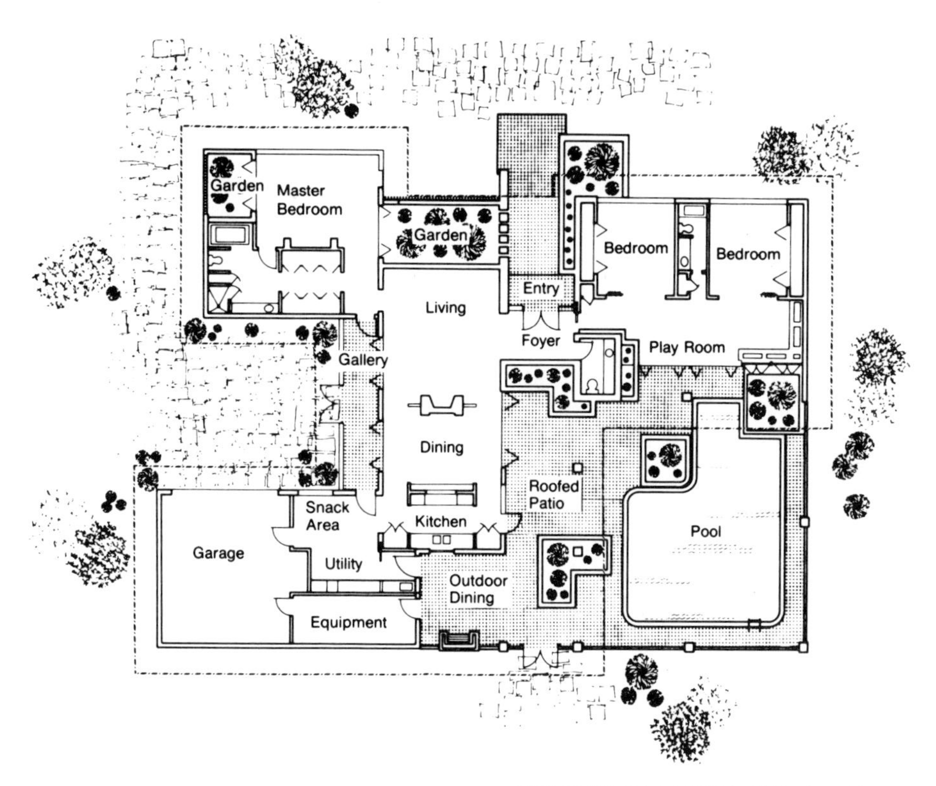

WESTINGHOUSE ELECTRIC CORPORATION
Floor plan shows how breeze is allowed to sweep over pool area, entering house to cool it.

nique reduces heat movement through the walls, ceiling, and glass. Compared with prevalent construction practices, Electra III reduces energy losses by 72 percent.

Most air-conditioned homes operate by means of a thermostat that turns the air-conditioner on when a preset temperature is reached indoors. Electra III's thermostat works in conjunction with a special energy-saving device called an "enthalpy controller," which checks outdoor temperature and humidity first.

If the air outside is below a certain temperature and humidity, fans suck it inside to cool the house. The energy-consuming compressor, if unneeded, is not turned on; cooling is essentially free.

Solar Collection Area A solar collector area 53 feet square is mounted on the roof of the house to deliver energy for the domestic hot water supply. In addition, the heat pump helps heat the hot water.

When in operation, the heat pump uses a compressor to heat domestic water. Heat given off is ordinarily lost in a conventional heating system. Compressor heat is saved by attaching a heat exchanger to the heat pump. This heat is in turn delivered to the hot water system, reducing the amount of electricity required to keep the water hot.

The swimming pool in the Electra III house is heated strictly by energy from the sun. The pool collector unit uses nine black plastic panels which supply 100 percent of its heat.

The house has three bedrooms, 2½ baths, a play room, a large living and dining area, a two-car garage, and an equipment room. It also has a covered entranceway and foyer, a roofed gallery outside the living and dining areas, and a snack area, a utility space and outdoor dining area next to the kitchen. The pool and patio area is fully screened.

The total annual electrical power consumption of the house is 29,270 kilowatt hours. Compared to similar houses with the same floor plan, Electra III saves 54 percent in energy consumption.

The free solar energy used in the pool is not included in that calculation. If pool heating is included, the total energy savings comes to 72 percent!

Maximum demand for electric power in the entire house system is 8.3 kilowatts. That compares with a peak load of 21.3 kilowatts for a comparable non-energy-conscious home. This is a reduction of 61 percent for the solar/heat pump/energy-saving house.

THE COMPLETE SOLAR MOBILE HOME

A 60-foot mobile home can be turned into a solar house, using space heating, hot water heating, and air-conditioning all provided by the sun's heat.

In an experiment by General Electric Space Division, nineteen General Electric solar collectors 3 feet by 8 feet were added to an already built mobile home. The installation provides 80 percent of the heating and cooling needed, and 95 percent of the hot water requirements.

The remodeling consists of added insulation, pipes, and pumps, none of which are in the living area of the 12-by-60-foot home.

Collector panels are composed of a black, heat-absorbing metal surface covered by two rigid sheets of clear plastic.

Solar heat trapped in the collector is transferred to water, then pumped to an absorption-type heating/cooling system. Additional heat goes to a 400-gallon insulated storage tank.

The tank saves up enough surplus heat to run the heating/cooling system on most cloudy days. When too much heavy weather occurs, a gas-fired boiler is used to back up the solar equipment.

The entire system operates automatically. The main goal was to make the mobile home operated by solar energy livable.

10

Adding Solar Heat: "Retrofitting"

The most practical way to acquire a solar house is to build it from scratch. Adding on a solar heating system to an existing non-solar heating system is called "retrofitting." Retrofitting can run into structural and design snags because solar components sometimes cannot be fitted into or onto existing dwellings without considerable remodeling.

Yet it is possible to remodel almost any existing house so that it can utilize solar space heating, solar cooling, solar hot water heating, and solar pool heating. The cost, however, in some cases may well prove exorbitant.

In certain exceptional situations, you may be able to do the job yourself. However, it is highly unlikely that you possess enough engineering, building, and technical skills for the job of retrofitting a house for a total solar energy system—that is, one including solar space heating, cooling, and domestic hot water.

It is, however, likely that you may be able to put in relatively simple systems like pool heating equipment, and even simple domestic hot water equipment at some sites.

HOW ONE MAN RETROFITTED HIS HOME

A qualified engineer recently did a complete job on his Livermore, California, home with the help of several friends. The remodeling was a project for a master's degree thesis. His purpose was to show that a house heated by conventional means could be turned into a solar-heated home.

He brought the job in at the astonishingly low figure of $1,500 for materials. All the materials were available at the local building supply store.

He worked from scratch on the solar collectors, designing them and building them himself. They were constructed in 4-by-8-foot rectangles with 3½-inches of fiberglass insulating the bottom. The absorber plates were made of corrugated aluminum painted black.

Each collector was covered with glass and sealed tight. It weighed only 4 pounds per square foot, a weight easily carried by the existing roof structure.

For the piping system, conventional plumbing was used to carry the heated water to a pair of 500-gallon tanks in a specially constructed and insulated "tank room." In one of the 500-gallon tanks was added a specially constructed heat exchanger made out of copper coils connected to the water pipe. In the second 500-gallon tank a smaller, 42-gallon galvanized-iron water tank was immersed.

The water starts out by being pumped to the roof of the house by a small electric pump. It flows down over the corrugated black aluminum, becoming heated by the greenhouse effect of the solar collector. It flows down by gravity to the tank room, where it enters the heat exchanger in the first 500-gallon tank. As it passes through the coils inside the tank, heat is imparted to the tank water. The heated water then continues from that tank to the second tank, where it enters the 42-gallon tank. There it releases even more heat to the second water supply. This second 500-gallon tank serves as a stabilizer, containing a leveling amount of heat.

From the second tank, water can be drawn out into a conventional hot water tank for use in the house's domestic hot water system. Water in the 42-gallon stabilizer tank then is pumped out of the second 500-gallon tank to be pushed up onto the roof for another cycle of heat-gathering.

Meanwhile, the two heated 500-gallon tanks are warming the air that surrounds them inside the specially insulated tank room. This air usually measures about 140°, at least for ten months of the year. Only for two months does the temperature drop to 90°.

When needed, this heated air is conveyed by ductwork into an insulated control room equipped with a fan and a conventional gas-run backup furnace. If its temperature exceeds 68°, which is

ALUMINUM ASSOCIATION

While not actually a retrofit, this traditional Cape Cod colonial shows how front of house gives no hint of solar collectors (which are mounted in rear). Positioned for southern exposure, collectors take in energy for heating *and* cooling. Note big upper deck for lazy-day sunbathing.

the desired temperature of the house's interior, the fan blows the air throughout the rooms. If the temperature drops below 68°, a thermostat kicks in a conventional backup gas furnace which adds enough heat to satisfy the system.

The preheated water for the domestic hot water system is almost always maintained at about 140°.

The 500-gallon tanks were made out of corrugated sheet metal storm-drain pipes. Sheet metal tops and bottoms were welded on. The conventional hot water heater tank, the 42-gallon tank, and the copper coils were bought at a building supply house, as were the pipes.

The solar collectors are mounted on a gentle 20° slant, facing the south. In the Livermore area there are sometimes rains, frosts, freezes, and fogs from the San Francisco Bay. The system delivers a large proportion of heat during an ordinary year. One solar collector delivers about 40 gallons of water at 140° constantly throughout the year. In all, there are eight collectors.

Natural gas is inexpensive in the area. Heating the house for a year runs only about $150. In ten years the retrofitted solar system will pay for itself.

A BIG-CITY BROWNSTONE RETROFIT

Solar energy systems are not confined to the suburbs or the rolling countryside. Many are set up in the middle of heavily populated metropolitan areas. It is not always possible for a building in a crowded city to use solar energy because high-rises nearby may cut off the sun, but many solar systems do flourish in any city.

One retrofitted hot water system in a New York brownstone cost the owners only $3,000—excluding labor, which was performed by friends.

The system involves copper panels and connections to the copper hot water tank already located in the basement. Although tall apartment houses were nearby, the owners of the brownstone found that none cast shadows on their selected panel site. The city district in which the brownstone is located is protected by zoning laws which forbid the construction of any more apartment houses.

The collector area of the brownstone consists of five panels of copper plate. Each panel measures 3 by 6½ feet—an area of 19½ square feet, with a total of 97½ square feet.

Each panel's absorber plate is painted black and covered with glass. Copper tubing runs vertically through the system to carry the fluid medium.

The system works well enough most of the time to sustain the domestic hot water supply at 120° F. When the sun fails to deliver enough heat, the gas-fired hot water system already in the house starts up and adds enough heat to reach 120° F. Then the gas cuts out and lets the sun do the work.

Figures show that the yearly gas bill, which ran normally at $900, has been cut by 60 to 80 percent. The system will pay for itself in three to five years. Because the installation of the system is considered "home improvement," it is depreciable and becomes a tax benefit.

One interesting wrinkle in the addition of this hot water system is that it has been found to cut down on the cost of summer air-conditioning. How? With the conventional hot water heater on all the time, the basement was kept so hot that the air-conditioner had to run a lot more than it does with solar heat!

A GREENWICH VILLAGE SOLAR SYSTEM

A cooperative apartment in New York's picturesque Greenwich Village was retrofitted to provide domestic hot water heating for the entire building.

The system was installed when the Housing Development Administration gave the owners a grant to repair the building under ordinary rehabilitation laws. The building had been condemned by the city of New York.

The heart of the solar energy system is a large area of copper flat-plate collectors located on the roof of the building. The collection area totals 600 square feet.

The system is a fluid-medium type, which pumps solar-heated fluid into the basement, where a heat exchanger in the large storage tank imparts the sun's heat to the water.

The system works so well it saves fuel costs totaling between $4,000 and $5,000 every year.

Work on the remodeling project, including the solar energy system, was performed by members of a community group. These amateur do-it-yourselfers trained themselves in carpentry and construction work in order to accomplish the renovation and reha-

SOLARON CORPORATION

Although a sight like this is a rarity today, many owners of brownstones and high-rises are bringing in solar heating equipment to cut down the cost of running urban apartments.

bilitation. The members were paid $3 an hour, with each working at least forty hours a week. Each worker put in one eight-hour day per week gratis. That unpaid day represented the worker's investment in his own apartment.

Called a "sweat equity" program, the project cost only about half what it would have under ordinary circumstances.

IS IT WORTH IT TO RETROFIT?

The fact that some peope are able to retrofit existing homes and buildings with successful solar heating systems does not necessarily mean that you can make a retrofit pay off in your own home. There are many factors to be considered.

Except for the California house described at the beginning of the chapter, the solar systems mentioned above are strictly domestic hot water systems.

If you plan to add a domestic hot water system, you can probably do the job with a minimum of construction problems, plumbing hassles, and added storage room in the basement.

By consulting Chapter Five on hot water systems, you can ascertain the amount of solar collection area you need to mount on the roof.

If your hot water tank is large enough, you can use it for auxiliary hot water and add another tank for solar storage. If your hot water tank is worn out or not in good condition, you can replace it with a larger tank containing both heat exchanger and heating coil—a two-in-one tank to serve both storage and backup system.

If you want to add only a swimming pool heating system, you should be able to plan it without making any major changes in the existing house. You can erect the solar collector plates separately from the house if you wish, and simply connect them to the filter system of the pool.

A review of Chapter Six on swimming pool heating installations will give you an idea of what you need to do.

A RETROFITTED SPACE HEATING SYSTEM

The installation of solar energy equipment to heat and/or cool an existing residence is a complex and expensive operation.

In fact, only by careful analysis can you ascertain whether or not it is feasible, whether or not it is economically sensible, and whether or not it is in fact practicable.

First of all, your geographical location is important in determining what kind of space system you want: a space heating system; a space cooling system; or a combination space heating/cooling system.

In certain areas of the country, a solar space heating system is adequate for comfortable living. Even though some days in the summer are uncomfortably hot, they can be ignored in consideration of the expense of installation and operation of an air-conditioning system for such a short time.

GRUMMAN CORPORATION

House in wooded area can be retrofitted for solar heat, but position of trees must be carefully studied to select proper part of roof for panels.

In other areas, a space cooling system is adequate for comfortable living. The relatively few cool days in the winter can be taken care of by auxiliary heating equipment.

It is entirely unacceptable to let a solar energy system languish for months on end without affording any help to the cooling or heating of your home. The matter is discussed in Chapter Eight, with reference to specific localities. For further information, refer to the monthly tables of cooling and heating in Chapter Fourteen.

For the majority of geographical areas, of course, a combination heating and cooling solar system which in turn supplies most of the domestic hot water supply can be a great aid in cutting down fuel costs.

Once you have decided what kind of heating, cooling, or heating/cooling system you want, you must then analyze your own residence and property to see if retrofitting is feasible.

The Solar Collector Subsystem Of primary consideration to any retrofit job is the placement and installation of solar collectors.

Siting. The physical design, construction, and material properties of your roof are of primary importance to the positioning of solar collectors.

(1) The pitch of the roof is quite probably several degrees off the optimum for your geographical position, which is approximately the latitude of your area. Situating the collector on a custom-built ramp or platform can solve this problem.

(2) The structural rigidity of the roof may present a problem. Will the roof hold the solar collectors? What kind of collectors are needed? How much do they weigh?

(3) The position of the roof in relation to the sun is a factor of great importance. The part of the roof on which the solar collectors are to be installed must have at least six hours of clear sunlight each day. Any less than that will produce inefficient results. Think of trees, hilly slopes, buildings, and so on.

(4) The siting of the collectors must be analyzed to be sure they do not present hazards to power wires, telephone wires, or neighbors. For example, if the reflection from the collectors shines directly in someone's window in the morning or afternoon, you must get assurances that there will be no hard feelings or lawsuits.

Optional siting. If the roof proves unacceptable as a mount for solar collectors, you can always improvise other sitings.

For example, you can mount them on an existent fence or wall that faces to the south.

If you have a garage detached from the house and not shadowed by it, you can mount the collectors on its roof slope.

You can even mount collectors on a south wall, keeping them vertical, with some loss in efficiency, or letting them slope at the proper inclination by spreading out into the yard.

The Solar Storage Subsystem Another important consideration in retrofitting is the installation of the storage tank or storage pit.

In many cases, the addition of a storage tank for a fluid-medium collector system presents no problem. The addition of an extra tank in the basement or cellar can be achieved without total dislocation and with a minimum of pipe installation.

However, it is possible that the house does not have a cellar or basement. If the existing hot water system and furnace is in a crowded space which cannot be expanded without excessive structural changes, there is always the possibility that you can use part of your garage.

If that is not possible, you may be able to bury the storage subsystem in the ground near the house.

Air medium. A collector system of the air-medium type presents more of a problem in the storage subsystem than in the collector subsystem.

The problem is obvious. Most air-medium systems store heat in pits of crushed rock. The large space needed for these rock pits is usually to be found under a house in the basement area.

If your cellar is already crowded, you may be able to locate the rock pit outside by burying it in the yard. Ducts connected to the rock pit will carry in hot air and pipe out warmed air on demand.

No matter which kind of collector system you install, you can use the heating and/or cooling system as a backup for the new solar energy system.

The Solar Distribution Subsystem A crucial factor in retrofitting is the type of existing heating and/or cooling system already in the house.

Forced-air system. The most efficient and most easily adaptable system of heating and cooling in the home is the forced-air system. This system delivers warmed air to every room of a house through a ductwork in walls and floors, moving colder air back to be expelled, filtered, and/or reheated.

(Photo: John M. Miller)
GARDEN WAY LABORATORIES

This two-hundred-year-old home is retrofitted with solar collectors, mounted on sloping roof. This violates ordinary siting rules by facing east, with others on opposite side facing west.

Solar energy can be used to heat and cool older, modest homes as well as new, expensive models. This retrofitted house is in El Cajon, California.

ALUMINUM ASSOCIATION

Because the typical solar space heating and space cooling system recovers stored heat either directly through air pumped over heated rocks or through air passing over heated coils of hot water, the forced-air ductwork can usually be connected just as it is to a newly installed solar collector and storage system.

However, if the existing forced-air system was installed initially only for heating and not for heating *and* air-conditioning, the size of the ducts may not be big enough to carry a solar heating supply.

If the forced-air system was installed for heating and cooling, the ductwork is adequate for a solar energy system and can immediately be connected to the new installation.

In the event that the ductwork is too small, you will have to consider having it replaced with larger ducts. A heating and cooling engineer can give you advice on this.

Hot water radiant system. If the existing system is a hot water baseboard radiation system, the solar energy system cannot be used without increasing the existing radiation area by about three to four times.

Structurally, this is a rather unattractive prospect. Also, the design of the rooms will be changed considerably.

There is one solution, however, which may be a bit costly, but which does solve the problem. By adding small fan coils in the baseboard units in order to boost their output, a baseboard system can be adapted over to solar.

Heat pumping heating/cooling system. If you plan to install a solar system in conjunction with a heat pump system, you need only hook up the collector subsystem with the heat pump for an excellent and effective augmentation of the heat pump's performance.

PREPARING THE HOUSE FOR RETROFITTING

With considerations of solar collection area, storage area, and distribution area out of the way, you must test your house for its tightness and ability to prevent energy loss.

If there are multiple air leaks, the addition of solar collectors will be a waste of time. The efficiency of solar energy can be reduced in half by careless weatherstripping, air leaks, and insufficient insulation.

To prepare any house for solar retrofitting, it is mandatory that the house be as airtight as possible, with proper "breathing" holes to prevent condensation.

Making a house energy-fit is not easy. However, in these days of high fuel costs, even a house that is not using solar energy should be carefully and effectively sealed against heat leaks.

Because this subject is such a complex one, an entire chapter is concerned with the subject. See Chapter Eleven.

Even after studying and analyzing all these considerations, it is sometimes not easy to decide whether or not to retrofit to solar energy. You should really discuss the situation with a building contractor who has had experience in the field of solar retrofitting. You should also ask him for names of satisfied customers so you can talk to them too.

Then you should really hire an experienced engineer to check out the contractor's claims. In addition, you should have your lawyer read the contract to determine exactly what construction and installation costs are covered and what warranties on performance and equipment are included.

A SPECIALIST IN RETROFITTING

International Solarthermics Corporation of Nederland, Colorado, specializes in retrofitting installations. Because of the problems encountered in trying to mount collectors on inadequate or badly pitched roof slopes, the company has come up with a controversial type of installation.

ISC models three sizes of a small back yard collector which stores solar heat in from 25,000 to 45,000 pounds of crushed rock housed in a small A-frame shed. The solar collectors are attached to one side of the shed.

For delivery, warm air from the heated rock in the shed is pumped into the ductwork of a conventional home heating system, with the home's initial system itself doing backup work.

The simplicity and easy installation of this back yard system makes its addition to a three-bedroom house under $6,000. The company requires 18 inches of fiberglass batts, or the equivalent, in ceilings of the solar-heated house. With such proper insulation, the back yard solar system will provide up to 90 percent of the average home's heating needs.

Climate, size, location of house, and size of collector shed all contribute to the final amount of heat derived and the efficiency of the system.

DO-IT-YOURSELF RETROFITTING

Another type of retrofitting that up to now has a limited appeal is the do-it-yourself project for solar energy installations.

At the present time, however, several firms are putting easy-to-do plans on the market for anybody to buy and use.

One of them is the Hadley Solar Energy Company, in Wilmington, Delaware. For $7 you get plans for a 4-by-8-foot solar collector that even a tyro can build for around $200 or less.

It was designed by the man who helped the University of Delaware construct its experimental solar energy home.

The do-it-yourself collector can be made to handle either air or water. The designer says that it will heat a well-insulated room 15 feet square.

Ironically, he got his ideas for the collector out of an old Department of Agriculture pamphlet—published in 1936!

ADD ON A SOLAR ROOM

What started out as a solar-heated greenhouse extension for far-north climates has turned out to be a specially designed add-on room not for plants but for people.

The solar room can be added onto any existing structure and will help heat a portion of the house as well. Best of all, the designer has drawn plans so that almost anyone with a modicum of ability at home repairs and carpentry can do the job himself.

With the plans, which cost about $10, any contractor can easily put on the room.

The designer includes several different-sized rooms in the set of plans. One is 12 feet long and the other 16. The addition of the solar room will save you some of your fuel costs for the year.

In New York, for example, with 4,871 degree days per year on the average, and about 57 percent sunshine during the heating season, the addition of the room can potentially save 32 percent of fuel bills.

In Burlington, Vermont, with 7,865 degree days, and sunshine only 41 percent of the heating season, the potential savings is 8 percent.

GARDEN WAY LABORATORIES
Solar room can be easily added to any existing structure. Do-it-yourself plans list materials, tools, and show how to put it all together.

GARDEN WAY LABORATORIES
Solar room has been added to old farmhouse. Windows must be unobstructed by trees or other obstacles.

The plans are specific and easy to follow. In addition, the solar room uses various types of energy-saving controls, in addition to the glass "wall" and the insulation.

For example, there is one sensor and two room-to-room fans, one installed at ceiling height to blow warm air into the house from the solar room, and the other at floor-level height to pull cool house air into the solar room for heating and then return it to the house.

The sensor is the brain of the system. It senses room temperature against house temperature. When the solar room is hotter than the house, the sensor activates a fan that blows the hot air into the house and reduces fuel consumption. If the house is warmer than the solar room, the fan remains off.

Construction materials include framing timber, plywood sheathing, concrete block piers, conventional wood flooring, insulation, roofing, wiring, and a special heat-storage niche in the room itself for the storage tank.

The list of materials is complete, as are instruction on how to put it all together. The materials are easy to get.

Offered by Garden Way Publishing Company of Charlotte, Vermont, each set of plans includes drawings and instructions for a solar room added on to the side of a house, or for a free-standing solar-heated greenhouse.

It can be sunroom, solarium, kitchen, study, bedroom, greenhouse, or solar room—whatever you want it to be.

11

Preparation for Solar Heat

A vital step in the preparation of an already existing house for a solar heating system is to make it as energy-efficient as possible.

That means sealing up all the heat leaks that have been around for many years but which haven't been noticed or fixed. It means using every method possible to cut down on the amount of heat used to keep you comfortable. It also means using the dozens of ways known to energy experts to cut down on heating and cooling expenditures. In fact, even if you do not plan to use solar energy in your home, you should think about making it as energy-tight as possible in order to save money on your fuel bills.

The best way to conserve heating energy—and cooling energy as well—is to have as energy-tight a house as possible. There are four specific ways of making your house snug: by insulation; by putting in storm windows and doors; by weatherstripping; and by caulking.

TYPES OF INSULATION

Proper insulation in your house can save up to 30 percent of your fuel bills. The beauty of insulation is that in an unfinished attic you can probably do the job yourself.

There are several different types of insulation material available: mineral wool (including fiberglass, rock wool, and mineral fiber), cellulose, and urea formaldehyde foam.

(1) *Mineral wool* is made from inorganic glass, rock, or slag, and is the most versatile and widely used of insulation materials. It comes in blankets, batts, pouring wool, or blowing wool.

- *Blankets* are rolls of insulation, 16 and 24 inches wide, some containing vapor barriers and some not. If the insulation has no vapor barrier, it is called "unfaced."
- *Batts* are similar to blankets, but cut into 4-foot or 8-foot lengths.
- *Pouring wool* is loose-fill insulation meant to be poured into unfinished attic floors.
- *Blowing wool* is loose-fill insulation meant to be blown in place by a contractor. Pneumatic equipment pumps the insulation into ceilings, attic floors, or walls through small holes drilled in the outside surface.

(2) *Cellulose*, made from newspapers and other paper scrap, is loose organic insulation meant to be blown or poured in place.

(3) *Urea formaldehyde* is a plastic formulation foamed in place at the building site. This is contract work. It is not recommended for attic floors.

HOW THICK IS INSULATION?

Insulation value is rated in R numbers, "R" standing for *resistance* to heat movement through the insulation. The R number is the actual resistance value; the thickness of the insulation product varies with the type of material.

For example, R-19 and R-22 mineral wool batts and blankets are 5 to 7 inches thick. R-11 and R-13 blankets are 3 to 3½ inches thick. Batts of higher resistance, up to R-30, are available, although they are sold only on special order.

You can buy mineral wool batts and blankets through your lumber dealer or insulation contractor.

WHERE TO INSULATE

The places where you will probably be able to add insulation in an existing house are limited to three main areas: attic floors; unfinished walls; and floors above cold space.

(Photo: Heilpern)
NATIONAL MINERAL WOOL INSULATION ASSOCIATION, INC.
Two layers of R-19 mineral wool batts double heat-retaining shield in ceiling of house with unfinished attic.

Attic Floor Because an open attic floor is easily accessible, adding insulation there is the easiest way to begin tightening up your house. R-30 or R-38 insulation is the best to use in the floor. Although it is thick, a large amount of attic floor insulation will partially compensate for poorly insulated walls in the rest of the house. If your home is heated, air-conditioned, or both, you will need at least that much insulation to keep the house warm or cool. Also, you may be in a climate that is cooler or warmer than average. And don't forget, energy costs are going to continue to rise.

If you're buttoning up your dwelling for retrofitting to solar energy, you'll need that much to help preserve the precious energy from the sun.

In the event that your attic floor is not insulated at all, install R-30 or R-38 mineral wool. For colder climates, use R-38. Incidentally, all R numbers are additive: two R-19s make an R-38; R-19 and R-11 equal R-30.

You can also get the proper R value with blown or poured mineral wool, or with a combination of blankets and loose wool.

If your attic floor already has some insulation—6 inches or less—add R-19 or R-11 to what is already there. Just lay mineral wool on the top of any existing material.

How to insulate an attic floor. If there is no insulation in place, lay batts between the joists. On the floor they do not have to be stapled in place. Just be sure the vapor barrier faces *down.*

If you are adding to insulation already there, simply place a layer of batts or blankets on top of the old material. In this case, the new insulation should *not* have a vapor barrier. Buy unfaced insulation. If you cannot get unfaced insulation, remove the vapor barrier or slash it freely with a knife and then install the insulation with the slashed surface down.

You can also pour wool into the space between the joists. Simply open up the bag, dump it out, and level it with a rake or short board.

Walls If the wall space you want to insulate is accessible, install R-11 or R-13 mineral wool blankets and then cover the insulation with an interior finish material like Sheetrock or paneling. To insulate finished walls, call an insulation contractor and have him blow the insulation in.

To insulate an unfinished wall, fit the end of a mineral wool blanket against the top stud or piece of framing. Staple the flanges of the insulation to the face of the stud, keeping the staples about 8 inches apart. With foil-faced blankets, staple the flanges to the *sides* of the studs to create an air space above the vapor barrier. This allows the heat-reflective value of the foil to be achieved under the finish surface applied over it.

Fit the blankets tightly against the studs at the bottom. If more than one piece of blanket is used in the same space, butt the ends tightly together. The vapor barrier must face the side of the wall that is heated in the winter.

If the stud spaces are narrower than the normal 16 or 24 inches, cut the insulation about 1 inch wider than the space to be filled. Staple the remaining flange, then pull the vapor barrier on the other side to its stud and staple the barrier in position.

It is actually possible to insulate a wall using unfaced blankets. Afterward apply a separate vapor barrier. Use 2-mil or thicker polyethylene sheeting or foil-backed gypsum board. Keep the polyethylene taut as you staple it in place.

(Photo: Heilpern)
NATIONAL MINERAL WOOL INSULATION ASSOCIATION, INC.
High degree of thermal protection is achieved with R-19 batts between studs in wall built with 2-by-6 timber.

As for pipes, ducts, and electrical boxes, install insulation behind them. You can even hand pack these spaces with loose wool from a blanket. Just pull pieces of mineral wool out of the packing.

In any unfinished wall, there are always cracks and narrow spaces left around a window frame. Remove loose wool from a blanket, and stuff it by hand into the cracks. Then staple a vapor barrier into position.

How to insulate masonry walls. It is possible to insulate masonry walls, basement walls, and the like by fastening furring strips vertically in place. Anchor them 16 or 24 inches on center—measured

(Photo: Heilpern)
NATIONAL MINERAL WOOL INSULATION ASSOCIATION, INC.
Small space between window and rough framing can be packed by hand with pieces of insulation taken from scrap of mineral wool batt.

from the center of one strip to the center of the other. Furring strips can be attached by masonry screws or bolts. Apply insulation blankets between the furring strips as if they were wall studs.

Floors To insulate floors above cold spaces, R-19 is recommended for all homes in the northern half of the country, and for homes in the southern half that are heated by oil, electricity, or solar energy.

Push batts or blankets between the floor joists from below, with the vapor barriers *up*. To hold up the insulation as you work, run baling wire back and forth in intersecting Xs, attached to nails about 2 feet apart at the bottom of the joists.

Or staple chicken wire to the joists to hold up the blankets. Heavy tape can be used as support, too, stapled to the bottom of the joists. At the outside of the floor area, cut pieces of blanket to size and fit them along the sill with the vapor barriers in.

VENTILATING ATTICS AND CRAWL SPACES

In both winter and summer, it is necessary to ventilate above the insulated floor in an attic or crawl space. In winter the open vents let moisture vapor escape. In summer, the moving air lessens the build-up of attic heat.

Provide at least two vent openings, located so that air can flow in one and out the other. A combination of vents at the eaves and at the gable ends is better than just gable vents. A combination of eaves vents and continuous ridge venting at the peak of the roof is best of all.

For Gable Vents Only Each 600 square feet of attic area insulated from the rest of the house with a vapor barrier needs two vents: a 1-square-foot inlet, and a 1-square-foot outlet.

Each 300 square feet of attic area insulated from the rest of the house *without* a vapor barrier needs two vents: a 1-square-foot inlet and a 1-square-foot outlet.

In other words, you need twice as much venting for insulation without vapor barrier as insulation with it.

For a Combination of Eaves and Gable Vents Each 600 square feet of attic area insulation from the rest of the house *without* a vapor barrier needs at least two vents: at least a 1-square-foot inlet and a 1-square-foot outlet, *with half or more* of the vent area at the top of the gables and the balance at the eaves.

The same figures apply to a combination of ridge vents and eaves vents.

WINDOWS AND DOORS

Once you have buttoned up your house with the proper insulation, you should move on to your doors and windows. Next to leaky walls and ceilings, most heat loss occurs through glass areas.

Glass is a highly heat-conductive material. Doors and windows by their very nature have cracks around them, even when closed. Door frames and window frames aren't always tightly sealed to the wall either. Air passes through these leaky joints.

There are three basic steps to take to tighten door and window openings: protect them with storm windows and doors; weatherstrip each; and caulk around the framework.

Storm Windows and Doors The addition of storm windows and storm doors can cut household heat losses in half. In the summer, of course, it keeps out heat gain.

If the windows and doors are already double-glazed—with two panes of glass—they can be made even more leakproof by the addition of storm windows or doors.

Triple-glazing—insulating glass plus a storm window—is even more effective than double-glazing, and is often used in extremely cold climates.

Depending on the type of storm window or door, a skilled home handyman can probably install one easily. If not, a contractor will do the job at a nominal fee.

You can make your own storm window or storm door by using plastic sheeting. Simply tape or tack the plastic to the inside of the window or the glass portion of the door. It is an inexpensive method and can be used with good results by people who rent their homes.

Clear plastic "storm window" can be installed inside regular window. Kit comes with plastic "frame" into which sheet of acrylic can be inserted for winter and removed during summer.

PLASKOLITE, INC.

Do-it-yourself kits are available for different kinds of windows. In the more fancy types, the plastic sheet is inserted each season into a plastic frame which remains attached to the wall all during the year.

Weatherstripping Most weatherstripping is simply installed and can be done by almost any do-it-yourselfer. There is a variety of different types of weatherstripping. They are made out of felt strips, foam rubber, flexible vinyl, spring bronze, and other metals and plastics.

Each type of weatherstripping material comes in a kit marketed at a hardware or building supply store. Instruction sheets come with the material, detailing very simple and easily understood directions.

The object of weatherstripping is to fill up the holes that naturally exist around a window between it and the window frame or around a door and the door frame.

In a house with an indoor–outdoor temperature difference of 35° F. (for example, 68° inside and 33° outside) a quarter-inch crack under a 3-foot-wide attic door costs $4.50 per winter in wasted fuel, according to the National Bureau of Standards in Washington, D.C.

Windows. Weatherstrip window sash on both sides and at top and bottom. In a double-hung window, don't miss the meeting rail where the top and bottom sash come together.

Doors Weatherstrip entrance doors at the top, at the bottom, where the latch meets, and at the hinge end. Special types of weatherstripping are available for the bottom inside of an entrance door—a flap that closes down when the door is shut to keep out the air blowing under the door between it and the sill.

Caulking Caulking, like weatherstripping, is a do-it-yourself job. You can buy caulking guns and caulking cylinders at hardware or building supply stores. When you exhaust one cylinder, you throw the empty away and buy another for insertion in the metal frame.

To caulk, press the trigger with your fingers and squeeze out a thin bead of caulking compound onto the crack you have found, or where you suspect a crack might be.

If the crack is too wide for a thin bead of caulk, you can use cord-type caulking. Press the cordage into place with your fingers.

If even the cordage won't fill the cracks, you can pull mineral wool out of insulation batts and stuff that into the holes before applying cord and/or caulking.

SAVING ENERGY WITH CARPET

With the house all buttoned up through insulation, weather-stripping, caulking, and storm windows and doors, there are several more simple methods to cut down on heat loss in the home.

Savings of up to 13 percent in annual heating bills can be achieved by the use of carpet in the home, a study by the Georgia Institute of Technology shows.

In addition to cutting down on energy consumption, carpet's soft pile reduces "floor fatigue." Note carpet on the walls to cut down on noise in work area.

CARPET AND RUG INSTITUTE

CARPET AND RUG INSTITUTE

Insulation of printed carpet on floor and solid-colored carpet panels on walls cuts energy consumption in heating and cooling.

The figure is based on the location, shape, and size of the home, local fuel rates, and seasonal degree days. In extreme climates the dollar savings is significant.

The tests show that combinations of carpet and carpet underlayment cut down heat loss through the floor by as much as 72 percent when installed on an uninsulated concrete slab. A like combination will reduce floor heat loss by as much as 54 percent on an uninsulated wood floor over a conventional ventilated crawl space.

The insulation values of the carpets in the test are additive for any combination of samples. A carpet with an R value of 1.3 and an underlayment pad with an R value of 1.6 will yield an overall R value of 2.9, according to the Thermal Subcommittee of the Carpet and Rug Institute, which initiated the tests.

It is the pile construction of carpet that provides its insulation potential. Air spaces or pockets between the fibers hold the warm air in and keep it from escaping.

CARPET AND RUG INSTITUTE

Even kitchen can be carpeted to cut down on energy losses. Carpet reduces incidence of slips and cushions falls to prevent injuries.

In summer months, carpet can be used to keep a house cooler by preventing the movement of heat from outside to inside through the floor. This cooling factor can cut down on air-conditioning costs the same way insulation can cut down on heating bills in the winter.

In some cases carpet has been used on walls to help insulate them. On a wall installation, the insulation value of the carpet is the same as it is on the floor.

SAVING ENERGY WITH DRAPES

The use of drapes in front of windows helps cut down on the amount of heat that escapes through the glass. The use of drapes can also help the homeowner adjust the window area as the sun changes position. For a north window, the closing of the drapes will effectively insulate the window area and prevent heat from escaping through the glass.

The space between the drapes and the glass acts as an insulation barrier. For that reason be sure to pull the drapes all the way closed for the maximum effect.

SAVING ENERGY WITH WINDOW SHADES

Window shades can save up to 15 percent of the costs of heating and cooling, other tests show. Under a grant from the Window Shade Manufacturers Association, the Illinois Institute of Technology in Chicago conducted a study which found that a drawn roller shade can prevent from 24 to 31 percent of the heat loss through the window. In summer, a sunlit window with a drawn shade admits 44 to 54 percent less total heat than an unshaded window.

A house with a 15 percent window area, according to the study, will reduce its heat loss by approximately 8 percent in the winter by the use of shades. In the summer, the reduction in energy required for cooling is more than 20 percent.

INSULATE YOUR WATER HEATER . . .

Whether you use solar energy or not, you can save a lot of energy loss by insulating your water heater. It is a do-it-yourself project that almost any handyman can tackle.

There are two types of heater insulation kits on the market. One is made for gas-fired hot water tanks, and the other for electric.

Specific instructions come with the kits.

The basic steps are the same. First you wrap the tank with a fiberglass blanket which has a vinyl surface on the outside, in order to find out where to cut the insulation. Mark the top and bottom of the tank, along with the pilot light (if it's a gas heater) and drain valve.

Cut the insulation to the right length with a knife or scissors and notch openings for the pilot light and drain valve.

Attach the fiberglass with double-stick tape to the tank. Then use vinyl tape around the outside.

. . . AND YOUR DUCTWORK

Kits also are available for you to insulate the hot-air and air-conditioning ducts. You can use any kind of 2-inch blanket material and wrap and tape the insulation to the ducts.

12

Solar's Challenge to Architecture

Buildings of all kinds—including single-family residential homes—use at least one-third of the energy that is consumed in the United States. Experts agree that 60 percent of this energy could be saved if all new buildings were designed to be energy-efficient. In fact, 30 percent of this total could be saved if existing structures were adapted for such efficiency.

Within fifteen years' time, such design changes would save the equivalent of 12½ million barrels of oil per day, or 4½ trillion a year. That amount is more than twice the current level of oil imports.

Most important in this consideration of energy-efficient design is the relationship between architecture and energy. Energy conservation by way of enlightened design is a primary weapon in the war against energy waste.

This kind of conservation is a necessary complement to the strategy of increasing the use of the sun's free supply of energy.

And the challenge to the architect is obvious: he must design energy-conserving houses without sacrificing any of the living standards Americans have come to enjoy.

Because windows are big wasters of heat, the whole concept of enormous window space must be reconsidered. Big windows everywhere may become a thing of the past in residential architecture.

The sight of shading or screening devices to protect homes from too much sun in the summer and to let in light and energy in the winter may soon become a familiar sight.

The old-fashioned overhang may stage a comeback. It is actually a most intelligent response to the conservation-oriented home. The amount of window space will have to be geared to the light-receiving capabilities of the house as much as to the esthetic taste of the designer and the big-window bias of the homeowner.

But esthetics is not the primary problem.

THE HOME THAT BLENDS IN

One of the main reasons American homes have used up so much energy in the past thirty years is the fact that money-conscious builders have put more emphasis on uniformity of design than on architecture that is compatible with the natural earth rhythms, the prevailing breezes, the movement of the sun, and the general lay of the land.

We should once again let our homes adjust themselves to the environment in which they are placed. Perhaps we can return to the days when windows could be opened at will rather than interminably sealed shut. Perhaps we can return to an environment oriented to the air and the sunshine, to an environment in which everyday people and their life purposes are in closer harmony.

Some architects and engineers are doing something about it. Numbers of experimental energy-saving houses, in addition to solar-heated structures, have been built within the past few years.

From a study of them a number of energy-saving features can be observed. By combining as many features as possible that save heating and cooling energy, the architect can put together the perfect house to support the solar energy collectors that will run it.

This ideal energy-conserving house can then be heated by solar energy, without allowing most of the precious heat to go out through cracks, through badly placed windows, or through heat-wasting escapeways caused by careless design.

POSITION OF THE IDEAL SOLAR HOUSE

The orientation of the solar house is of primary importance. It must be built on a site where a maximum of sunshine falls, at least in a portion equivalent to the needs of the structure.

Generally speaking, a collection area roughly equal to half the living area of the house is needed to catch the sun's light.

That means the southern portion of the lot must be free of high obstructions, dense foliage, or man-made structures that screen out the sunlight.

In areas where heating is a primary consideration, the presence of trees may not be of too much concern. The absence of leaves during the heating season is nature's way of contributing to man's home heating systems.

In heavily populated areas, with small lot space, the height of a neighboring structure to the south must be taken into consideration. The house can be placed further north on the same site, or it may be necessary to elevate it.

In densely populated areas like cities, the concept of solar energy becomes more complicated. In many instances, solar collectors have helped achieve heat savings, but usually the amount is nowhere near that of heat savings on a single-residence property.

Because positioning of the house is so important, the architect must keep in mind the sun's elevation throughout the year.

LAYOUT OF ROOMS

To utilize solar energy to the utmost, the actual layout of a house should take into consideration the following facts:

- By placing the sleeping area—bedrooms and baths—either on the north side of the house or on the east, these rooms will receive sun in the early morning only or not at all.
- By placing the living areas—family room and living room—on the south and west side of the house, these rooms will get sunshine during the day and in the late afternoon, leaving that portion of the house still warm when the sun sets.

Meanwhile, the sleeping area has cooled off in the summertime for resting comfort. In the winter, the arrangement offers maximum heat for family activity in areas during the afternoon and evening hours.

THE SHAPE OF THE HOUSE

Even the shape of the house, oddly enough, contributes toward its energy-conserving or energy-wasting potential.

A square house actually wastes less heat than a rectangular one. Therefore, the closer to a true square the shape of the house is, the less heating and cooling energy it will lose when it is built.

The reason for this oddity is obvious. It is at the perimeter of the house—that is, the outside walls—where the heat is actually conducted outward (or inward in summer).

The house with the least lengthy perimeter, then, is the house that is structurally the most energy-tight. Of all the conventional shapes of houses—squares, rectangles, Ls, Ts, and so on—the square has the smallest area-to-perimeter ratio.

FLOOR PLAN OF HOUSE

There are two general types of floor plans in residential architecture. Most familiar is the "closed" plan in which the living area is broken up into separate rooms. This conventional type of living arrangement favors privacy of individuals, but cuts down on the overall efficiency of the heating and cooling system. To satisfy the various compartmented areas, the system must deliver heat and cool air to each room separately.

An "open" plan means a minimum of interior partitions. This uncluttered mode of design allows heat and light to flood a greater area of the house than it would in one of comparable size with more interior partitions.

There is a good reason for this type of design. Heat rises, moving from hot areas to cold. With an aperture between two rooms, warm air moves to colder areas. The movement of the air tends to waste heat, and the circulation of the warmed air is impeded by the barriers.

For summer cooling, cooler air can move more easily from one area to another to cool the entire area quicker and at less expense.

CONFIGURATION OF THE HOUSE

In a two-story house, an open system tends to work like the chimney of a fireplace.

In summer, cool air is introduced below either through windows or doors on the first floor, or through the cooling duct system, is

(Photo: R. Ward)
AMERICAN PLYWOOD ASSOCIATION
Partitions and ceilings are minimized in open house design. High vent-type windows pull out warmed air in summer. Warm air is recirculated by blowers into main area in winter.

drawn up through the house, and is exhausted through open windows in the second story.

In winter, the house can be made to work like a reverse chimney. Heat from the sun is collected directly through the windows, and through solar collectors via the distribution subsystem. From there, with the help of fans, the heat at the top of the "chimney" can be redistributed throughout the house by means of a blower and duct system. The house itself acts like a giant backup solar heat collector.

COPPER DEVELOPMENT ASSOCIATION
Energy-saving features include: sun's heat which is trapped in lower-story greenhouse-type area; house designed in energy-saving square configuration; south-facing windows and balcony door set back under overhangs.

PLACEMENT OF CHIMNEY

One most important consideration in the energy-saving house is the placement of the wood-burning fireplace. Since time immemorial, the chimney has doubled as an outside wall of the structure.

For energy-saving measures, however, in a solar-heated house the chimney is placed against an inside wall or is freestanding. This allows the heat that radiates out from the chimney to warm the air

(Photo: R. Ward)
AMERICAN PLYWOOD ASSOCIATION
Two-story energy-saving house has wood stove and chimney installed in middle. Heat escaping up chimney radiates outward and adds to warmth of house.

inside the house rather than escape outside and warm the outside air.

Even in summer, in a cooling situation, an inside chimney is preferable to an outside one. With the outside chimney, heat from the outer area creates an updraft in the chimney itself, sucking out cold air from below which would be better left inside the house.

WINDOWS AND OVERHANGS

The positioning of windows is extremely important. Generally speaking, the largest windows should be placed on the south side of the house, including portions of the east and the west for early morning and late afternoon insolation.

Normal 2-foot overhangs are doubled to 4 feet in energy-efficient house. Note small row of windows beneath overhang. WESTINGHOUSE ELECTRIC CORPORATION

COPPER DEVELOPMENT ASSOCIATION
Double-paned door on right and windows at center and left reach from floor to ceiling in energy-saving house. Space between panes insulates and cuts down energy loss.

On the north side of the structure, windows should be small and inconspicuous. The small size cuts down on heat loss in winter, and allows it to be used in the summer as a kind of vent system, drawing hot air out of the house but not creating such a large draft that the cooled air is pulled along with it.

The position of the sun in the summer is almost directly overhead; in the winter it is quite a bit lower to the horizon.

In the summertime, an overhang to cut off the sun's hot rays from directly overhead can protect the larger areas of all south-facing windows. In the wintertime, the overhang will not affect the reception of sun's rays lower on the horizon when they are needed.

Windows and doors should be carefully placed to offer sources of natural light and ventilation, and should be treated especially with caulking and weatherstripping so as not to allow excessive winter heat loss and summer heat gain.

DOUBLE DOORS AND AIR LOCKS

Two extremely effective methods of conserving energy have been borrowed by architects and engineers from other technologies.

One of these is the double door, which prevents the escape of desired air into an undesired area. A double door is simply two doors hinged to a door frame rather than one. One of the doubles

Clear all-heart redwood creates formal setting and adds insulative value to room. Vertical blinds control excess sunlight in summer and open for light in winter.

CALIFORNIA REDWOOD ASSOCIATION

opens outward, the other opens inward. When the doors are closed, an extra thickness of dead air produces excellent insulation that keeps heat from moving in either direction.

Another design innovation is the air lock entryway. Borrowed from outer and inner space technology, the air lock is a simple and effective means of conserving vital heated air or cooled air.

The old-fashioned entryway was essentially half an air lock. One outer door opened into a vestibule separated from the main part of the house only by open space. The outer door was usually the only one opening into the vestibule.

The air lock is a vestibule with a second door opening out into the rest of the house. With the outer door open, some air from the outside environment enters. When the door is closed, the unwanted air is cut off. When the inner door is opened, little of the unwanted environment is allowed to enter.

CONSTRUCTION MATERIALS

Throughout an energy-saving house, high-quality and energy-conserving materials must be teamed up with imaginative design ideas.

The use of wood in an energy-conserving structure has been proved to be more efficient than most kinds of materials, with the exception of insulation material, which is usually used as an adjunct rather than as a construction material alone.

Tests conducted at Arizona State University compared the fuel consumption of two structures identical in size. One was built with wood, the other with masonry. Both structures were insulated and both were exposed at the same time to identical weather conditions and controlled interior temperatures.

The results? The wood house used 23 percent less energy than the masonry house for winter heating and 30 percent less for summer cooling.

Actually, wood is an excellent insulator, simply because of the way it grows. A report from the American Plywood Association tells why. Made up of cells that contain millions of tiny air spaces, wood insulates four times better than cinder block, six times better than brick, fifteen times better than concrete or stone, four-hundred times better than steel, and one thousand seven hundred seventy times better than aluminum.

CALIFORNIA REDWOOD ASSOCIATION
Superior insulation values in this house come from wood-frame construction, wood-paneled walls, wood ceilings, wood stairs, and even wood banisters. Carpet on landing adds to energy-saving motif.

Wood-frame construction, wood-paneled walls and ceilings, and even window blinds not only offer excellent insulation, but can help decorate any room. In addition, they help insulate against noise as well as heat loss.

A wood ceiling has many advantages. It can offer a dramatic change, it is relatively maintenance-free, it offers a natural counter-balance to white or light walls, and it is part of a trend toward increasing use of natural, long-lasting materials.

Overhead wood paneling provides an effective barrier against a room's rising heat supply which can be rapidly lost through an uninsulated ceiling. Wood also provides for sound absorption of much of the everyday noise a household generates.

CALIFORNIA REDWOOD ASSOCIATION

Ceiling paneling of sapwood-streaked redwood adds warmth not only in decorative effect but in insulative value as well. Energy-conserving room features drapes, Navaho blanket on wall, and double-paned sliding glass doors.

Actually wood in all its forms has self-insulating qualities, and the California Redwood Association finds an increasing number of architects, builders, and interior designers specifying redwood panels—not only for walls, but for ceilings, kitchen cupboards, and bathroom cabinets as well.

Even in bathrooms, redwood paneling is increasing in popularity in the mountain and western areas. Special finish—a satin-finish polyurethane varnish—is recommended to protect the wood from the steam and water in the bathroom. Additional paints, stains, and preservatives are not necessary because of redwood's own built-in preservative nature.

The fact that wood has such excellent insulation qualities does not rule out the use of other materials to produce the best kind of energy-conserving house.

Actually walls made out of concrete blocks and thick concrete aggregate are excellent retainers of heat rather than insulation barriers. The trick is to use them for their retentive qualities rather than for insulation.

For example, the early Southwest American Indians built houses out of adobe, a kind of brick made from local mud. But they built the walls a foot or more thick. Walls of that thickness keep cold out and warmth in during the winter, and warmth out and coolness inside during the summer.

How? Adobe's great heat-retentive qualities turn the wall into a heat storage zone. During the day, adobe absorbs heat as the sun beats down on the walls. At night the heat radiates out into the night, but some also comes into the living area.

By putting insulative barriers on the outside of the walls at night, the heat will radiate only inwardly.

Ordinary earth retains heat in the same manner. A house constructed on sloping land can be easily made heat-retentive by sinking one of the walls into the ground.

As the caveman knew, earth and rock retains heat long after the sun sinks out of sight, and gives up heat during the cooler hours of the day.

Adobe is not indigenous to the rest of the country, but rock is. By constructing walls that are thick enough, the use of rock can provide the same kind of heat-retentive qualities as adobe.

The same potential is true of thick concrete aggregate, or even concrete blocks that are thick enough. The use of concrete blocks, coupled with excavation and construction of half the house underground, can make up a cozy home.

Brick walls, built thick enough, can be used for the same purpose.

CONSTRUCTION OF HOUSE

In any energy-saving house, construction must be first-rate. All joints in the structure must be caulked to insure tight construction and to reduce any air leaks. Any chinks in construction which allow air to move unimpeded cost money in heating and cooling bills.

Insulation is essential in any economical, energy-wise construction. Careful attention must be given to both the type and the amount used. (See Chapter Eleven.)

The use of additional floor insulation can be provided by means of cushion-back vinyl flooring and carpeting.

Doors and Windows The same careful attention given to insulation and roofing must be applied to windows and doors as well. (See Chapter Eleven.)

If windows are wisely constructed, placed, and protected, there is no need to trade esthetics and natural light that is so conducive to good living for simple economy.

The optimum energy-saving house has windows and glass doors throughout with double panes separated by a half inch of dry air space. This space of dead air provides high insulation value.

The glass panes should be encased in specially treated wood frames that reduce heat and cold movement through the glass.

If sliding doors are used, they must be manufactured of tempered safety glass. They should be protected by the positioning of the porches, keeping heavy winds from damaging them.

Roof overhangs also can be used to shade them from excessive sunlight.

MECHANICAL SYSTEMS

The heating and cooling distribution unit of the future is, of course, the heat pump, which has been described in Chapter Four.

It is one of the most efficient types of equipment available for heating and cooling.

With a heat pump, even the heat generated by the operation of the pump can be used to heat the house. The result is that for every unit of energy used by the heat pump, two units of useful heat are returned to the house.

A system using an electric heat pump is just as effective for summer cooling as it is for winter heating, as explained in Chapter Nine.

FORCED-AIR HEATING AND COOLING

Coupled with the heat pump, of course, is the distribution subsystem. Although hot water distribution subsystems will work in the solar energy house, the most efficient is the forced-air system with ducts that carry warmed and cooled air wherever it is wanted.

One reason for its efficiency is its ability to reverse itself in the summer and provide cooled air rather than warmed air. With hot water and steam, an extra air-conditioning unit is necessary, operating at a higher rate of fuel usage.

The ductwork can be geared to economy and energy conservation as well. A fully insulated forced-air duct system can be located around the perimeter of the house so that any heat loss through the walls and the glass areas can be counteracted.

With ductwork placed not under the house, as in conventional design, but rather within the floor panels, the energy-saving house can cut down heat loss with the ducts actually radiating the lost heat up into the structure.

13

The Cost of Solar Heat

Operating a heating system run by solar energy is relatively inexpensive in comparison to operating a heating system run by the burning of fossil fuels. The initial investment required in the installation of a solar heating system, however, can run into thousands of dollars.

An average estimate is about $8,000 for a house with 1,800 square feet of living space. And, because the sun doesn't shine every day of the year throughout the entire country, a conventional backup system to make the solar system work during cloudy and stormy weather is needed. That costs about $1,500 to $2,000.

Because solar energy is free—no fuel is involved—the installation of solar heat will pay for itself after a while. Estimates on the length of pay-back time vary from several years to several decades.

You can look at it this way: Money invested in solar energy equipment will increase the value of your house. As long as you live in the house, you can enjoy the benefits of the improvement. And, in addition, you can save money on your yearly heating bills.

There is one other factor to consider: As the cost of fossil fuel—oil, gas, and coal—goes up, along with the cost of electric-resistance heat (since it is in large part generated by fossil fuel), the installation costs of solar heat will diminish in relation to the increased cost of fossil fuel operation.

HOW MUCH DOES SOLAR HEATING COST?

Today's solar heating system costs a lot of money to install because the solar industry has not yet developed mass-production methods that will eventually reduce costs. Why? Because there is not yet any great demand for solar heat.

This is seemingly a strange situation because it costs much less to *operate* a solar heating and cooling system than it does to operate a conventional heating and cooling system. Yet the initial outlay is considerable. And that cost varies a great deal from one part of the country to another.

HOW TO FIGURE THE COST OF SOLAR ENERGY

To find out if the installation of a solar heating system is practicable in your own situation, you should consider not only the size and location of your home, as well as the other factors discussed in Chapter Twelve, but also the cost of your existing heat requirements.

Four steps are necessary to determine the amount of savings you will effect against the amount of the cost of installation.

(1) Determine how much you spend on space heating each year at the present time.

(2) Find out how much you will save in operating expenses by the installation of solar heat.

(3) Determine as closely as you can the actual cost of installing a solar space heating system.

(4) Compare the amount of heating cost you save with the amount of installation money you pay.

The first step is the easiest. Just get out your check book and figure out how much you spent last year in heating bills.

You'll find that natural gas costs around $400 on an average; oil costs around $600 to $800 on the average; and electric-resistance heating is higher than that. Costs and requirements vary considerably throughout the country.

Don't forget that gas and oil prices are rising rapidly, so that your last year's bill will probably be exceeded next year simply through natural inflationary pressures.

How Much Money Will Solar Heat Save? Solar heat is practicable wherever winter sunshine is abundant, where heat requirements are high, and where fuel is expensive—in short, in areas like the North Central states, the Mountain states, and parts of New England.

In the sunny West, solar heat can be expected to provide about 60 to 80 percent of all energy needed to run a home (including hot water needs)—with space heating requirements almost 100 percent in some cases, particularly where the weather is warmer.

In the cloudy, overcast East, solar energy can supply only about 35 to 60 percent (including hot water needs)—with space heating requirements perhaps lower than that in some cases.

Insolation is the amount of sunshine available in any given area of the country. It varies from one part of the country to another. In Chapter Fourteen insolation has been measured in twelve "zones" of the United States, with the amount available for solar space heating and solar water heating noted for each month of the year. A study of those charts will give you an indication of the percentage of your space heating costs you can hope to save by means of a solar system.

The overall average of the amount of fuel you will save anywhere is about 75 percent, but of course every locality is different.

How Much Does It Cost To Install Solar Heat? Solar heating experts estimate that the installation of a solar space heating system costs about $4.50 per square foot of living space in a home. Naturally, this figure varies from one place to another, and even from one house to the next.

For a typical "average" house with 1,800 square feet of living space, the total cost of solar installation is about $8,000. That amount is derived from these working figures:

For an efficient space heating system, the average house needs from one-quarter to one-third of its living area duplicated in efficiently roof-mounted solar collectors. Using the one-third figure as optimum, the 1,800-square-foot house would need 600 square feet of collectors.

The cost of the average commercially available solar collector runs about $10 per square foot. Paying for the 600 square feet of collectors at this rate comes to about $6,000.

By the time you figure installation labor, the cost of the storage unit, and the cost of solar hardware such as pipes, blowers, pumps, sensors, and controls, you've added another $2,000, bringing the grand total to approximately $8,000.

Heating Cost vs. Solar Cost Let's take a hypothetical case. Please note that the figures used in this example are average ones only. Every individual case will be different, varying from one part of the country to the other, and depending on the number of degree days of each heating season and the cost of fuel in each area.

The idea is for you to substitute your own figures and work out your own estimate by means of the formula.

Suppose that you do have an average three-bedroom house with 1,800 square feet of living area that is heated by means of natural gas, the cheapest of home fuels. An average heating figure per year is $400, with variations for weather and cost.

With a solar energy system, you would save an average of 75 percent of your heating costs, as explained above—assuming, of course, that your saving is that overall hypothetical "average." There is really no way of telling exactly how much you will save until you have the system in operation. For the $400 heating figure, a 75 percent savings comes to $300 ($400 × .75 = $300).

As we've seen, the hypothetical average cost of solar installation is $8,000. If you pay for that installation in cash, your savings of $300 a year will pay for the installation cost in twenty-seven years. At the end of that time, you'll actually save $300 a month in heating bills.

It is highly unlikely that you will pay $8,000 in a lump sum. Assuming you take out a twenty-year loan of $8,000 at 8 percent interest, you will be paying $814.80 per year to pay it off. The interest rate is subject to variation, depending on where you live, what your bank charges, and other differences. To find out what your real cost for installation will be, substitute your own figures.

The solar heating system you have installed will save you $300 on the average, as we found above, if you use natural gas. If you pay $814.80 per year to pay your installation loan, you'll find it costs you $814.80 a year to enjoy a savings in operation expense of $300.

Until you pay off your loan in twenty years, you will actually be losing money—$418.80 a year—on solar heat, a total of $8,296 over a twenty-year period. However, once the equipment is paid for, you will begin to pay back $300 each year to the installation costs. In twenty-seven years you will stop paying for the installation costs and really *save* $300 a year!

Those figures assume that natural gas rates will not rise in twenty years. If they rise at a modest 5 percent each year, your $400 heating bill will be inflated to $620.53 in ten years. If they rise at 10 percent, your heating bill will be $1,037.49 in ten years!

If your home is heated by oil, your yearly cost is probably somewhere around $800, varying from place to place and company to company. The installation of solar heating will save you a hypothetical average of 75 percent, or $600. That's only a bit over $200 less than your investment.

Within eight years, if oil rises 5 percent each year, you'll be breaking even on your investment.

On the other hand, if your home is heated by electric-resistance heat, the cost runs about $1,200 a year, as a hypothetical average. The installation of solar heating will save you three-quarters of $1,200, or $900 a year. That's almost $100 more than your investment in the solar energy system costs.

With annual fuel savings of $300, you will pay back your $8,000 investment in twenty-seven years. With annual fuel savings of $900, you will pay it back in nine years.

HOW TO COLLECT MORE SUNSHINE

To make solar heating more competitive with oil and gas heating, engineers think a great deal of improvement can be made in the design and construction of flat-plate collectors.

About 300 BTUs of solar radiation per hour fall on a square foot of surface at midday in cloudless weather at a central latitude in the United States. The average solar collector now on the market can only capture about 120 BTUs of that amount. That's less than half!

Collection efficiency could almost be doubled, experts believe, by incorporating various improvements. For example, if glass covers are acid-etched to cut down on reflection, a great deal more sunlight will get through onto the absorber plate.

And again, if the black paint on the absorber plate is replaced by a black "selective coating" applied by chemical or electroplating techniques, less heat will be reradiated from the absorber plate.

Also, the creation of a vacuum inside the box will eliminate a great deal of heat loss, and consequently raise the efficiency of a collector by about 50 percent.

Owens-Illinois produces solar collectors based on the vacuum principle of design. Unfortunately, the price for their Sunpak runs about $20 to $25 per square foot!

However, the vacuum-type collector is a versatile development, one that can be used not only for winter heating, but for summer air-conditioning as well. Gas-fired cooling systems of the absorption type can be modified to run on water heated to 220° F. by the sun. Solar-heated water supplied by most air-medium or fluid-medium solar collectors is usually not quite that hot.

If space heating systems ever become commonplace, their cost will certainly drop, probably to well below the current $10 a square foot, according to most solar heating specialists. Good aluminum storm windows cost somewhere around $1.50 a square foot installed, a price considerably lower than when they were introduced some years ago.

In a solar industry grown to maturity, a decent flat-plate collector may cost no more than $4.50 a square foot. The completely installed system may not run much more than $7 a square foot.

Solar expert George O. G. Löf, director of the Solar Applications Laboratory at Colorado State University, and an early solar collector designer and innovator, points out that when gas and oil heating costs were in the $200 to $300 annual range, there was no reason for the public to use solar heat.

However, with yearly heating costs now rising to about $400 and more, and, in electrically heated houses in some localities even to about $2,000, a saving of up to 75 percent by the use of solar heating is "definitely attractive."

Löf predicts that in the next fifteen years, as conventional energy grows scarcer and more costly, there will be a dramatic increase in the number of solar-heated houses.

COST OF TYPICAL SPACE HEATING SYSTEMS

In Tucson, Arizona, which has one of the nation's best climates for the use of solar energy, Ernest Carreon, a builder, estimates that a sun-power heating system in a three- or four-bedroom house adds roughly $5 a square foot to the cost of the house.

He built a 1,200-square-foot home powered by a solar heating system; the cost of the house was $45,000. He estimates that it would have cost $39,000 or $40,000 if he had installed a conventional space heating system only.

His solar system will pay for itself through energy savings in eleven years at current electrical rates. At current natural gas rates, it will take sixty-two years.

In nearby Austin, Texas, the cost of installing a solar energy unit to heat and cool a 3,000-square-foot house is about $12,500, or $11,000 more than a conventional system burning fossil fuels, according to Dr. Gary Vliet, a University of Texas professor.

Much of this cost is in the water storage tanks holding 8,000 to 12,000 gallons of water, buried and insulated, that are needed to store heat for up to three consecutive cloudy, sunless days.

Dr. Vliet estimates that mass production and other factors can bring the cost down to $8,600 within several years.

THE HIGH COST OF SOLAR COLLECTORS

Much of the problem of high cost lies in the economics of solar collector manufacture. The materials alone for a typical air-medium collector cost over $3 per square foot—including sheet steel for the absorber plate, special insulation, heat-resistant gasketing materials, and high-strength tempered glass of the "low-iron" type that allows a very high percentage of sunlight to pass through.

By the time other costs are added in, plus profit for the manufacturer and distributor, the retail price for the collector alone is roughly $10 or more a square foot. Storage and installation costs bring the total price to $17 per square foot!

Not all collectors cost so much, although many are in that range.

Harry E. Thompson of suburban Washington, D.C., whose house is described in Chapter Two, believes in a simpler, more basic approach to solar heat. He uses collectors with single-pane glass covers in which water trickles down open grooves in sheets of black-painted corrugated aluminum. These collectors are retailed at a low cost—$4 a square foot—ready for mounting. That's quite a difference from $17.

Experts are split on Thomason's no-frills approach. Some authorities on solar energy believe his system is about the most workable one there is. Others claim that the use of open troughs causes water condensation on the inside of the glass cover, which in turn lowers the efficiency and makes the system less than effective for space heating in regions with very cold winters.

For mild climates, however, the system works well.

THE NOT-SO-HOT SOLAR COLLECTOR

One energy veteran, Hoyt C. Hottel of M.I.T., an authority on the flat-plate collector, points out that the average specimen made today has a very modest output—something between 100,000 to 250,000 BTUs per square foot per year.

Each square foot of solar collector surface saves the equivalent of 2.5 gallons of heating oil a year, he says, which is a savings of $1 with oil priced at 40¢ per gallon. To be price-competitive, Hottel points out, the solar collector system cannot cost more per square foot than ten times the yearly saving in fuel.

In our "average" house—1,800 square feet with 600 square feet of collectors—the cost of the installation of the system is about $8,000. The saving per year, according to the figures above, comes to $600. Ten times $600 is $6,000. Thus the installation price is not actually competitive—and would be only if the system came in at $6,000, two thousand dollars less.

To be competitive with oil heating, the price limit on solar equipment would have to be kept at a figure between $4 and $10 per square foot of collector surface. Many systems cost far more than that, particularly those used for space heating.

At present prices, an oil system is simply cheaper than a solar system, as is a natural gas system.

COMPETITIVE SYSTEMS

However, Hottel does believe that flat-plate collectors can save money for swimming pool heating. Because of the very low temperatures demanded of those heaters, the absorber plates never become hot enough to lose any significant degree of heat through reradiation back to the atmosphere.

Also, they do not require expensive glass covers, nor backing of thick insulation. Nor is pumping and heat storage necessary.

The total cost of the average pool installation is about $4 a square foot, with the collectors averaging only about $2.50 per square foot.

A hot water solar heating system by Sunworks costs about $1,000 to $1,600, and can pay for itself in less than ten years. The

Sunworks system works best in a house with an already installed electric hot water heater.

THE TROUBLE WITH SPACE HEATING

Space heating is a much more expensive proposition and one that is responsible for the high cost of flat-plate collectors. One major problem is the fact that the entire solar space heating system may sit idle for half the year, especially when the house doesn't need heating, but cooling.

As a general proposition, solar space heating systems now on the market are economical only if used as a secondary source of warmth.

For a building that must be warmed around the clock, a large expanse of collectors is needed. In northern states, where good sunlight is available only six hours a day on clear winter days, a house with 2,000 square feet of living space needs at least 500 square feet of collectors, although 650 is a more optimum figure. Space heating systems also require large heat-storage systems, which increase the cost up front.

It is particularly expensive to retrofit a space heating system into an existing home, though it can be done.

Two areas of the country where solar heat is sometimes the only answer to the energy problem are the Southwest and the West Coast. In many localities of these areas, new homes are being planned where natural gas hookups are no longer available or allowed, and in which there is no dependable supplier of heating oil or propane.

More and more builders in these areas are installing electric-resistance heating, which is very expensive. In situations where electricity costs more than 3½¢ a kilowatt hour (the equivalent of $1 a gallon for heating oil), a system of flat-plate solar collectors can definitely save money.

A SOLAR/ELECTRIC HYBRID SYSTEM

A new solar heating housing project in North Easton, Massachusetts, which is building 160 units all heated with solar energy, is using components supplied by Solaron Corporation of Denver, Colorado.

The system is a combination of forced-air heat from solar energy and electric-resistance heaters located in the hot-air ducts. The combination costs more than either an all-electric or an all-oil system. Each solar/electric system will cost about $3,500 apiece; during the life of a twenty-five-year mortgage, each owner will recoup his expense and start saving money.

Total cost to install *and* operate the system over twenty-five years will range from $335 to $516 a year. The cost of an oil heating system would range from $382 to $742, of a solar/oil combination system from $551 to $618. That's an annual saving of from $47 to $226 over oil heat and of $96 to $216 over solar/oil heat.

HOW TO SHOP FOR A SOLAR HEATING SYSTEM

Solar heat for the home—except for the houses of a handful of solar old-timers—is a new development. Many of the firms manufacturing hardware and installing it are novices at the trade.

If you are in the market for a solar system, you should be aware of a number of do's and don'ts before you go shopping and buying.

(1) If you finally find a system that looks good and that you are willing to take a chance on, don't sign a contract immediately. Ask the dealer or contractor for proof of performance. If he can't give you any, or seems evasive, hire your own engineering consultant to investigate the system. The money you pay the consultant will be well spent—even if he says the system will work.

(2) Insist on a warranty for the system and its installation. Examine the document carefully. Be sure you know the limitations of the warranty. Does it cover parts? Does it cover service? How long is the warranty good for?

(3) If you plan to put together a good system piecemeal by buying one part at a time, think twice. With solar systems so new, many of the bugs have not yet been ironed out, even by the professionals. It's a good idea not to try to "do it yourself" all the way with a complicated space heating system, unless you are a skilled technician. Kits for smaller systems are available—a one-room add-on, for example; or a swimming pool heating system. Be sure you can handle such jobs before you spend money on them.

(4) Ask the dealer for a list of previous purchasers of the equipment. When you get the list, go around and investigate the installa-

tions. Talk to the people who are using them. Find out how the systems work. If you are unable to do that, check with the local office of the Better Business Bureau to ascertain the dealer's credit rating.

(5) With any new and exciting development like solar energy, a number of fly-by-night operators will try to get in on the ground floor to take your money by false representation. Be particularly cautious about sellers of systems who only go by post office box numbers. Be sure you talk to the people you are buying from. Use normal caution with anyone you deal with.

HOW ABOUT A HOME-IMPROVEMENT LOAN?

According to information from the National Solar Heating and Cooling Information Center, interest in solar energy systems is growing so fast that many forward-looking lending institutions are beginning to give home-improvement loans and second mortgages for the installation of solar energy systems.

The situation varies from place to place. Some bankers are becoming aware that a house with a well-designed solar energy system is likely to increase in value as fossil fuel costs rise. Some bankers offer a reduction in interest rates for loans to people building energy-conserving homes—among them those heated and cooled by solar energy.

Other bankers are not so receptive to the idea. You'll just have to shop around to find the right one.

Don't forget, a properly designed system with a proven track record for saving fuel dollars will be a valuable feature in your home five or ten years from now. By then, solar homes may be selling at a premium.

HOW ABOUT TAX ADVANTAGES?

Certain states have passed legislation providing a property tax incentive to homeowners who install and use solar systems. As of this date, they are: Arizona, Colorado, Illinois, Indiana, Maryland, Montana, New Hampshire, North Dakota, Oregon, New Mexico, Texas, and South Dakota. Other states may soon follow.

In Congress, federal tax incentive bills are being considered and argued. Write your congressman to push them through.

HOW ABOUT GOVERNMENT GRANTS?

The Department of Housing and Urban Development recently provided grants of $400 each to 10,000 homeowners and builders in ten states for the installation of solar-heated hot-water systems in residences.

The grants covered about half the cost of the solar heaters, not including installation expenses. Ten states were selected to participate in the project, including Connecticut, Massachusetts, New Hampshire, Rhode Island, Vermont, Delaware, Maryland, New Jersey, Pennsylvania and Florida. They were selected because they were areas where electric heating bills were high last year or because they had shown an interest in encouraging the use of solar energy in residences.

The purpose of the grants was to show the feasibility of solar water heaters and to encourage their manufacture.

HOW ABOUT SOLAR HEAT DEVELOPMENTS?

A housing developer in North Lauderdale, Florida, recently offered an option of a $1,350 energy conservation package including a solar hot-water heater on the houses he was building.

Almost half of his buyers took the package, which featured special insulation and a wind-driven attic exhaust along with a solar heating unit.

Soon the homeowners were realizing savings of $30 a month on their electricity bills.

The solar energy heating device became standard equipment ir the building's new 170-home development, started one year later.

A homeowner who had moved there from Staten Island, New York, found he was saving up to $117 a month in heating bills. He had paid $157 a month for gas and electricity in New York, and only paid $58 in Florida. The solar unit alone cost him about $800.

In Washington, a Federal Housing Authority official said that as far as he knew, the Florida developer was the only one who included a solar energy device as a standard feature in an extensive development.

14

Zones of Insolation

Solar energy can be used to heat and cool a house effectively in almost every part of the country. Naturally, there is less solar energy available in certain regions than in others. In a locality with a high average number of cloudy days each year, the sun will obviously produce less heat.

To compensate, solar collectors should be larger, or there should be more of them to collect heat when the sun is shining. It is also a good idea to provide greater storage capacity for solar heat in an area where the sun does not shine on a high average of days.

This does not mean that solar energy is more feasible or more efficient a source of power in the Southwest than in the Northeast. It simply means that the system is different in each area.

In the Northeast, for example, fuel costs are much higher and the heating season is much longer. Both these facts make it economically desirable to use solar energy, making it at least as attractive economically as in the Southwest.

In the Southwest, the hot days of summer make it necessary to cool a house rather than heat it. Solar air-conditioning and cooling methods are more complicated than solar heating systems, and cooling costs more. Yet with the heating needs cut in the winter, the house in the Southwest can be cooled with solar energy at about the same cost.

Insolation is a word that refers to the total amount of solar radiation received by an object—for example, a house or a piece of ground. The amount of insolation varies throughout the United States. Climatic factors like temperature, cloud cover, humidity, and wind also effect the ways in which solar energy systems are planned.

A study of the insolation in the United States has been made to show how much sunlight can be expected in certain areas, how much energy can be supplied by solar collection to a home, what size of solar collectors is needed in each area, how much heating and cooling energy can be supplied by solar means, and so on.

For the purposes of locating your particular area, the country is divided up into twelve zones, called "solar climatic zones." The map shows the breakdown.

Since the zones do not always flow along state lines, the list on page 179 is necessary to show what and how many zones are included in each state.

The most important question for you to answer if you want to fit a solar energy system to your house or if you want to build a new solar energy home is the correct amount of insolation available to you in your geographical area.

Exhaustive studies have been made by teams of research scientists working for the Energy Research and Development Administration. The results of those studies show how much solar energy can be supplied for your home, how much collector area you need to catch the sun's energy, the average capacity of a storage tank, and other considerations.

Even the dimensions of representative solar collectors and the dimensions of representative cylindrical storage tanks have been determined.

To find the answers to all the above questions for the average 1,500-square-foot residential dwelling, all you have to do is consult the chart on page 180.

For example, if you live in Zone 7 you should plan to use solar collectors whose combined area measures about 500 square feet. You should plan to use a storage tank of 750-gallon capacity. You should expect 70 percent of your needs both in space cooling and heating to be supplied by the sun.

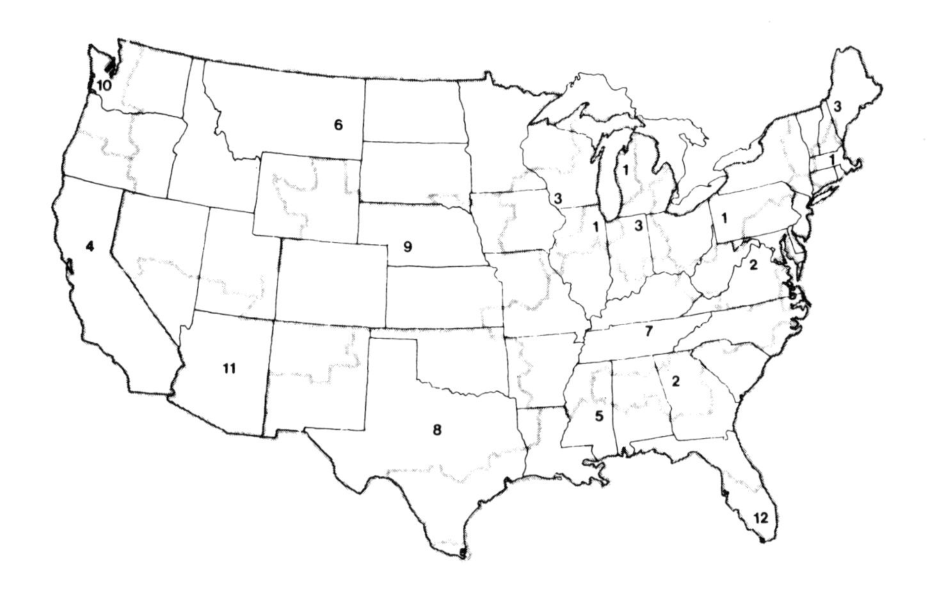

Breakdown of zones by state. Numbers in parentheses following each state indicate zones included in that state.

ALABAMA (2, 5, 7)
ARIZONA (11)
ARKANSAS (7, 8)
CALIFORNIA (4)
COLORADO (9)
CONNECTICUT (1, 2)
DELAWARE (2)
FLORIDA (5, 12)
GEORGIA (2, 5, 7)
IDAHO (6)
ILLINOIS (1, 3, 7)
INDIANA (1, 3, 7)
IOWA (3, 6, 9)
KANSAS (7, 9)
KENTUCKY (2, 7)
LOUISIANA (5, 8)
MAINE (3)
MARYLAND (2)
MASSACHUSETTS (3)
MICHIGAN (1, 3)
MINNESOTA (3, 6)
MISSISSIPPI (2, 5, 7)
MISSOURI (3, 7, 9)
MONTANA (6)
NEBRASKA (9)
NEVADA (6, 11)
NEW HAMPSHIRE (1, 3)
NEW JERSEY (2)
NEW MEXICO (8, 11)
NEW YORK (1, 2, 3)
NORTH CAROLINA (2, 5, 7)
NORTH DAKOTA (6)
OHIO (1, 2, 3, 7)
OKLAHOMA (7, 8)
OREGON (4, 6, 10)
PENNSYLVANIA (1, 2)
RHODE ISLAND (1)
SOUTH CAROLINA (5, 7)
SOUTH DAKOTA (6, 9)
TENNESSEE (7)
TEXAS (5, 8)
UTAH (6, 9, 11)
VERMONT (1, 3)
VIRGINIA (2, 5, 7)
WASHINGTON (6, 10)
WASHINGTON, D.C. (2)
WEST VIRGINIA (1, 2, 7)
WISCONSIN (3, 6)
WYOMING (6, 9)

Climatic Zone	Percent of Energy Supplied by Solar	Collector Area, Square Feet	Representative Collector Dimensions		Storage Tank Capacity, Gallons	Representative Cylindrical Storage Tank Dimensions	
			No. of 8-Ft. High Rows	Length of Each Row, Ft.		Diameter, Inches	Length, Inches
1	71	800	3	33	1,500	48	200
2	72	500	2	31	750	42	138
3	66	800	3	33	1,500	48	200
4	73	300	1	37.5	500	48	78
5	75	200	1	25	280	42	60
6	70	750	3	31	1,500	48	200
7	70	500	2	31	750	42	138
8	71	200	1	25	280	42	60
9	72	600	2	37.5	1,000	48	132
10	58	500	2	31	750	42	138
11	85	200	1	25	280	42	60
12*	85	45	1	5.5	80	20	63

* Includes only hot water needs.

Climatic zones	Energy savings, millions of BTUs	Comparison with oil		Comparison with electricity		
		Equivalent gallons	Dollar savings *	Equivalent kilowatt hours	Dollar savings at indicated cost	Cost per kWh
1	67.9	757	303	19,900	995	5¢
2	54.9	612	245	16,000	720	4.5¢
3	82.0	914	366	24,000	960	4.5¢
4	41.8	466	186	12,200	488	4¢
5	33.8	377	151	9,900	347	3.5¢
6	98.9	1,103	441	29,000	1,015	3.5¢
7	50.6	564	226	14,800	518	3.5¢
8	39.0	435	174	11,400	399	3.5¢
9	74.6	832	333	21,900	767	3.5¢
10	46.5	518	207	13,600	272	2¢
11	43.7	487	195	12,800	512	4¢
12	16.7	186	74	4,900	196	4¢

* 65% furnace efficiency at 40¢/gallon

In fact, you can even compute your savings not only in energy BTUs, but in gallons of oil, in kilowatt hours of electrical energy, and in dollars and cents based on set prices.

Note that 1 kilowatt hour is considered to be roughly equivalent of 3,412 BTUs.

Studies of the amount of solar energy available in BTUs were made for each month of the year in all the twelve geographical zones on the map.

The charts on the following pages give the findings, including the amount of solar energy available, the amount of solar energy collected, the amount of energy required for space heating, and the energy required for hot water heating.

For example, if you reside in Zone 5, you should use a collector of 200 square feet, and can expect to collect a little over 3 million BTUs in July, which is more than enough to supply energy for your hot water system. Obviously, there is no need for you to collect energy for heat at that period of the year.

In January, your solar collector should bring in 4.5 million BTUs, more than enough to supply your hot water needs, but not nearly enough to supply your space heating requirements.

This particular finding points up a need that you probably would not have anticipated. If you want to increase your solar heating capacity beyond what the chart shows you that you should get, you need only increase the total of your collector area.

However, increasing the number of your solar collectors is going to add to the cost of the installation. Not only that, if you decide to put up more collectors than those required to average out the year, you must also increase the size of your energy storage tank.

Solar heating engineers use a rule of thumb to the effect that the storage tank should be able to hold about 1½ gallons of water for every square foot of collector area.

For air systems that store heat in rocks, the storage volume of the rock should be about three times that of a water system.

In other words, 1 square foot of collector area should have 1,038 cubic inches of crushed rock available for heat storage. That equals more than half a cubic foot, which is 1,728 cubic inches.

The tables in the charts give the approximate cubic content in cylindrical inches.

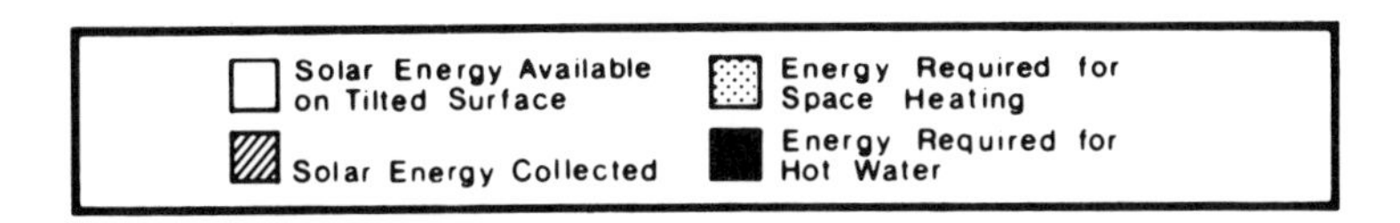

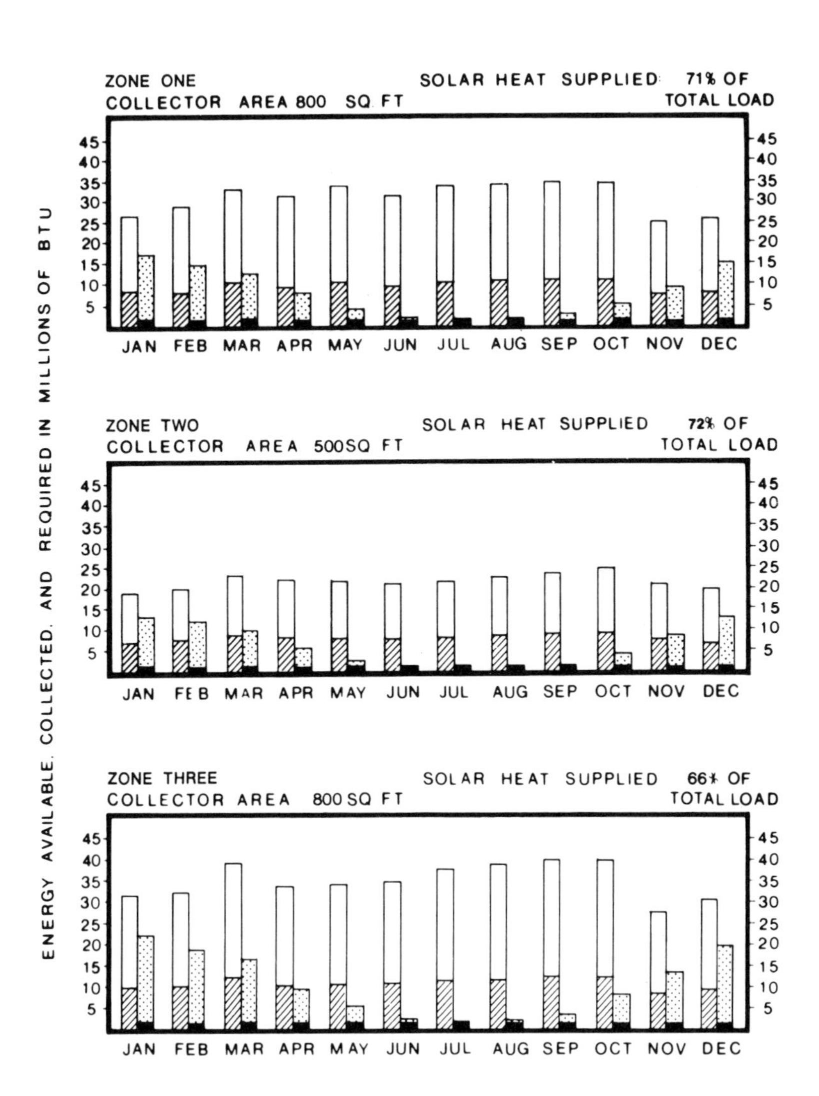

Energy Available, Collected and Required For a 1,500 Square Foot House Located in Climatic Zones 1-3.

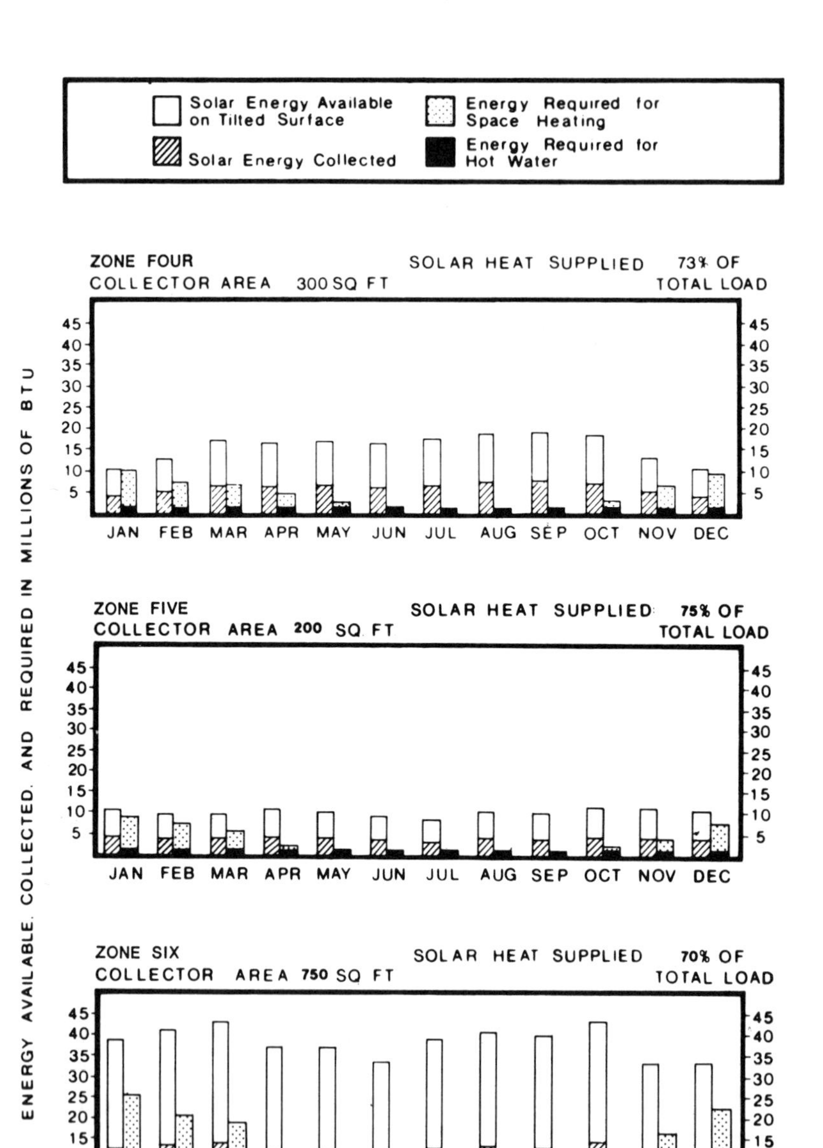

Energy Available, Collected and Required For a 1,500 Square Foot House Located in Climatic Zones 4-6.

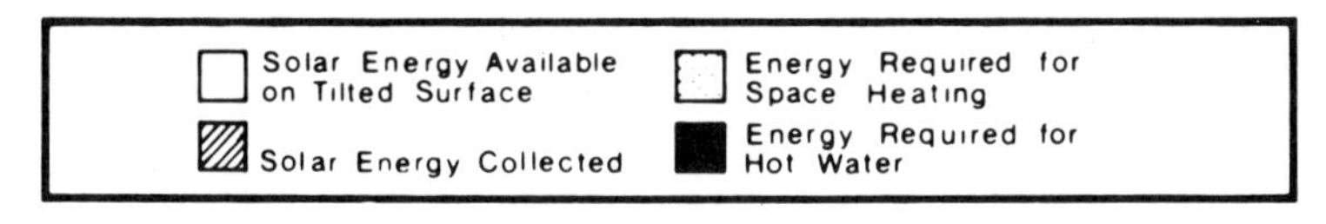

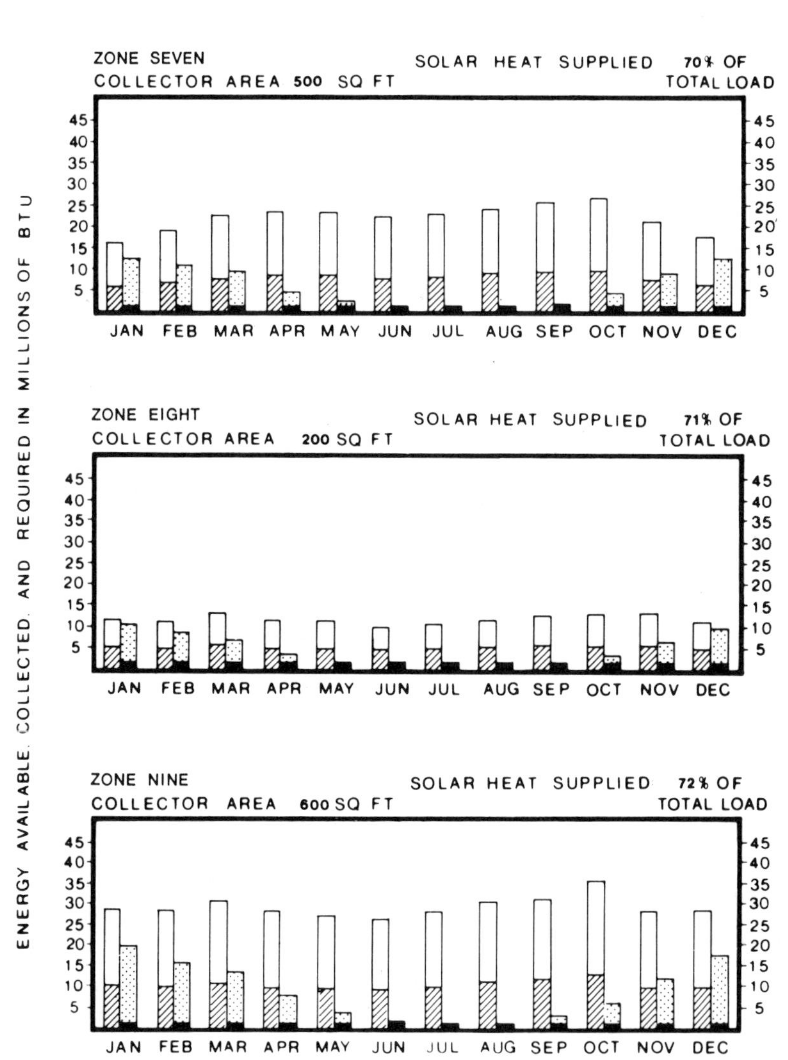

Energy Available, Collected and Required For a 1,500 Square Foot House Located in Climatic Zones 7-9.

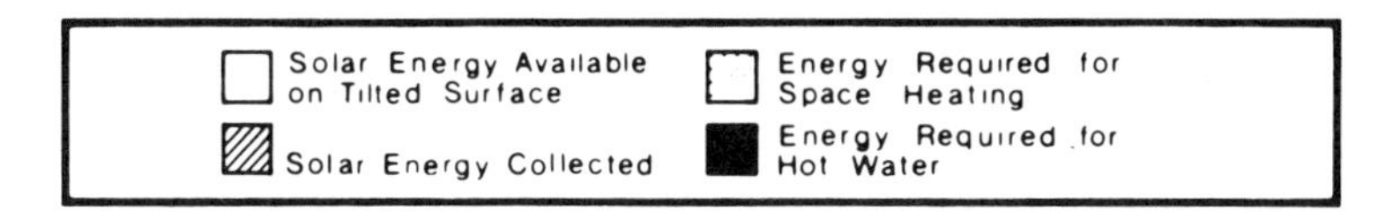

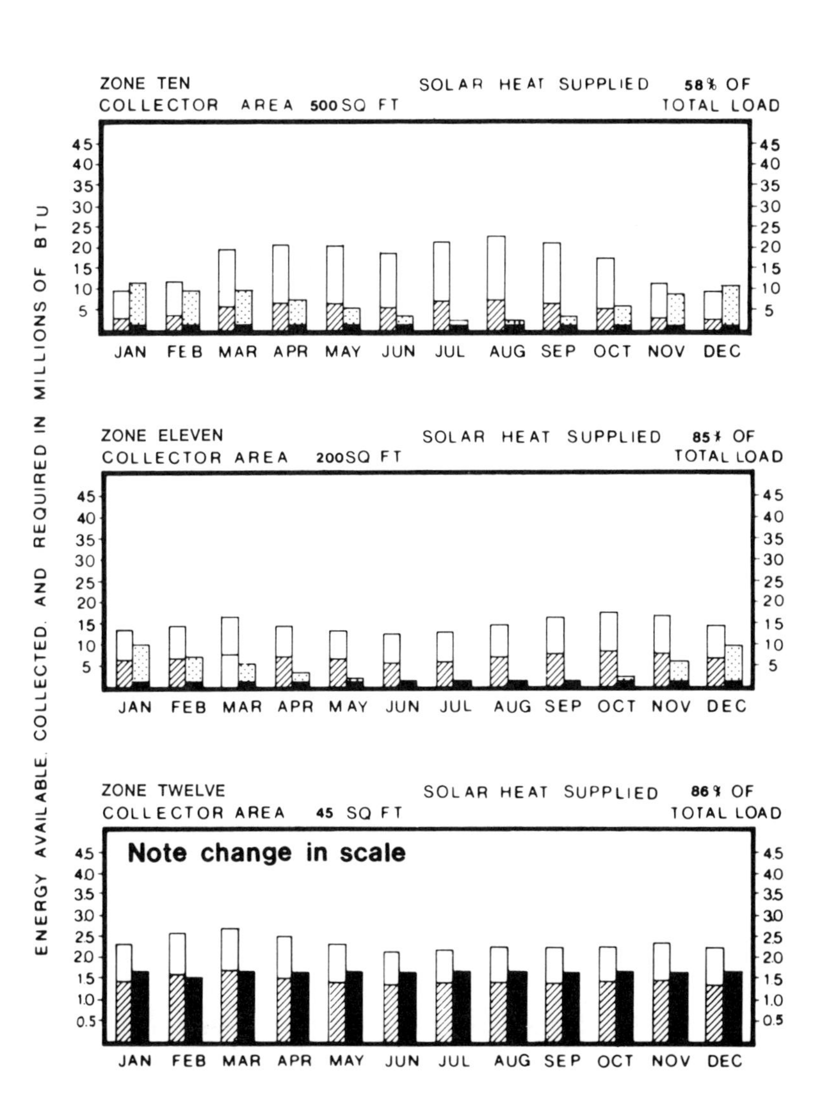

Energy Available, Collected and Required For a 1,500 Square Foot House Located in Climatic Zones 10-12.

15

Electric Power from the Sun

Oil and natural gas are fossil fuels burned to provide heat by combustion. The heat in turn runs machinery and warms houses.

Electricity is not a fuel, but a force of power used to run machinery or appliances operated by electric motors, or to impart space heat to buildings through resistance coils.

Fossil fuels provide heat to be used immediately and on the spot. Electric power can perform work far from its generated source.

Fossil fuels are burned to generate electric power for distribution to remote sites for use in heating, cooling, or running appliances. However, it is not necessary to use up prodigious amounts of fossil fuels to generate electric power. The sun can provide enough heat to do the job.

Not only that, but electricity can actually be stored for future use in batteries charged by the sun's light. These cells are similar to ordinary dry-cell and wet-cell batteries. They are called solar cells.

The development of solar-charged batteries and solar electric generators is in the works at the present time. The technology is known. Only the practical developments are needed.

HOW THE SUN'S RAYS GENERATE ELECTRIC POWER

The technology responsible for converting solar energy directly to electric power was developed in the early days of solar experimentation. In Chapter Two, J. A. Harrington was mentioned as the inventor of a solar machine that could supply electric power enough to illuminate the light bulbs in a mine.

Modern solar generators are based on the same technology. A steam turbine generates electricity by means of a conventional dynamo. The steam is provided by a boiler heated by a concentration of intense solar energy.

Mirrors or lenses are used to bring large amounts of sunlight onto a boiler surface to bring the water inside to a boil.

Ordinary flat-plate collectors cannot be used. They are able to produce temperatures only up to 200° F. However, a large group of mirrors or lenses can supply concentrated heat that reaches at least 1,000° F. and more.

A typical solar electric plant experimental station is located at the Energy Research and Development Administration's Sandia Laboratories near Albuquerque, New Mexico. There testing will soon go on to perfect methods of producing 10,000 kilowatts of electricity—enough to serve a town of about 10,000 people.

When the experimental plant is in operation, it will be the nation's first solar electric pilot plant.

The heart of the electric generating system is the lens or mirror that concentrates or reflects the sun's rays onto the receiver plate at the boiler. In the Sandia operation, almost 100 acres of land are used for a collector field. About 40 acres of that will be planted with 20-foot-square mirrors—320 of them. A 20-foot-square mirror has an area of 400 square feet. The cost of each mirror is about $28 per square foot, or roughly $11,200 per mirror.

Each mirror must be kept in position every second of the day so that the sun's rays will be reflected directly at the boiler or receiver plate. To keep each mirror tracking correctly, a clockwork mechanism moves it so that its angle of reflection is always accurate. Each mechanized mirror is called a heliostat. The electronic clockwork adds another $8,000 or so to the cost of each mirror.

MARTIN MARIETTA CORPORATION
Artist's concept of solar electric plant shows projected field of tracking mirrors—called heliostats—moved by clockwork to keep sun's rays focused on boiler in tower.

A 10,000-kilowatt generator field needs only about 320 20-foot-square mirrors; a 50,000 kilowatt generator field would need five times that! In the future, fields of 20,000 heliostats may not be uncommon.

The most of the solar electric generator is tied in with the heliostat components. Cost estimates vary, but one figure pegs the cost of each kilowatt produced at $8,500. A 10,000-kilowatt plant would cost $85,000,000! That's a cost of $8,500 for each person who uses the plant.

Other estimates are somewhat less, but not less enough to put the solar generating plant in competition with an oil- or coal-fired plant.

Of course, as technology improves, and the components of the system become mass-produced, the cost per unit will drop. But it will take a reduction to about $1,500 per kilowatt before the solar electric plant will pay for itself.

MARTIN MARIETTA CORPORATION

Photograph of scale-model steam generator cavity, which serves as focal point for sun's rays from heliostat field.

Air-cooled pilot plant at center of solar electric generating complex will tie in subsystems and produce power for surrounding communities.

MARTIN MARIETTA CORPORATION

As if money isn't enough of a problem, there is also the awkward fact that the sun does not produce any heat when it is behind a cloud, behind a mist, or even behind a dust storm.

Certain solar heat storage systems borrowed from early technology are being tested out now. One of them involves the storage of solar power in the form of heated water underground at the receiver (boiler) site.

So far a system to maintain the generating plant's electrical output for four to six hours has been developed. A larger storage capacity would push the cost beyond competitive means.

Once in operation, solar electric plants will be used most probably in the sunny Southwest, probably only to supplement conventional power sources. That means that the plants will be of inter-

MARTIN MARIETTA CORPORATION
Electric power is generated here by steam turbines. Excess solar energy is stored in ground underneath area.

mediate size, say from 50,000 to 250,000 kilowatts, rather than the 1,000,000 kilowatt coal-fired and nuclear plants now under construction.

Solar energy experts point out that the current economic problems are all part of the natural cycle of development. Many of them think that solar electric plants will become economically competitive with nuclear plants and some fossil-fueled plants between 1990 and 1995.

A top ERDA official says: "When solar power becomes economic, it will sell itself."

Solar electric plants will save fossil fuel, will reduce particle emission of the nuclear type, and provide a longer life for conventional equipment that will not have as heavy use.

Meanwhile, conventional energy prices keep going up all over the country. While they do so, solar energy is becoming more attractive to everybody.

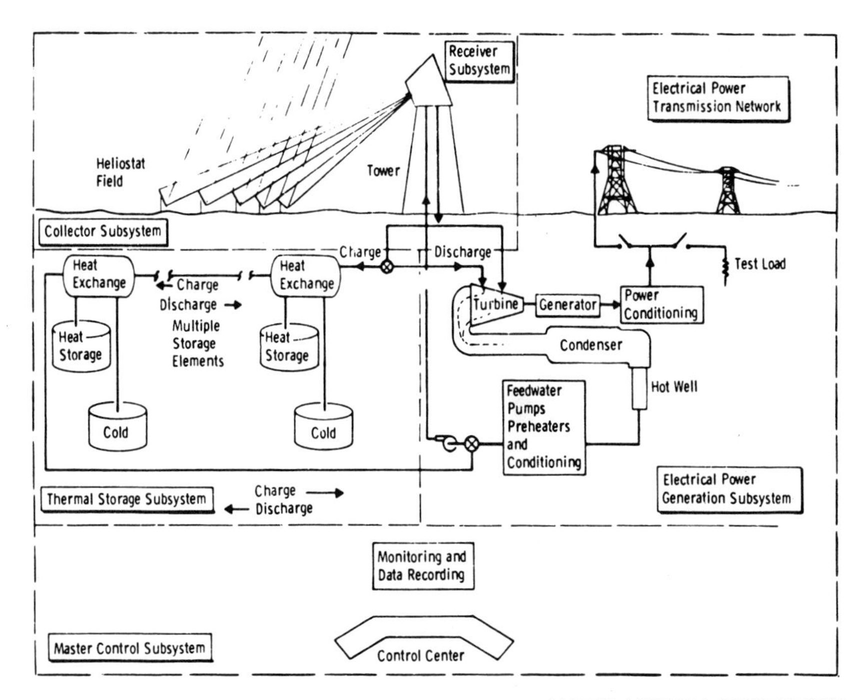

MARTIN MARIETTA CORPORATION
Schematic shows planned solar electric power plant capable of generating 10,000 kilowatts.

HOW ABOUT SOLAR BATTERIES?

The ability to convert sunlight to electric power by chemical means was known as far back as 1839, when Antoine Cesar Becquerel made his first discoveries in photochemistry.

In 1873, Willoughby Smith discovered that when the element selenium was struck by sunlight, its ability to conduct electricity increased.

The ability of selenium and other elements to react to sunlight eventually brought about the development of the selenium cell, which directly converted sunlight to electric energy that can be stored.

Later on, with the development of the silicon cell, that element became the primary metal for solar batteries. The silicon cell has a conversion efficiency of 11 percent, against about 7 for selenium.

LEFT

New silicon cells have developed even higher efficiencies—high enough, in space flight usage, to become standard equipment in ships.

The solar cell is called a photovoltaic cell—photo = light/voltaic = voltage. Scientists call photovoltaic cells "barrier layer" cells. The composition of the cell—layers of metal insulated by barriers—uses sunlight to create the flow of electric power.

The working purpose of the silicon cell is to produce a stored electric current. Alternate wafers of silicon, in crystal form, are placed—one with a negative charge, and the other with a positive charge—between insulating barriers.

The negative-charge silicon crystal wafer is produced by introducing a 5-valence element like arsenic onto the 4-valence silicon. This has the effect of fusing silicon's 4 electrons and leaving the extra arsenic electron free to move.

The positive-charge silicon wafer is produced by introducing a 3-valence element like boron onto the 4-valence silicon, fusing 3 of the silicon's 4 electrons, and leaving the fourth silicon electron free to move.

By interposing a semiconductor metal between these wafers—the barrier layer—positive and negative charges form on opposite sides of the barrier. By connecting these two opposite sides with a wire, an electric current flows from one charge to the other.

Almost no current flows when the cell is at rest. When light strikes it, however, the silicon reacts normally by increasing its conductivity. Thus the latent electric current begins to flow. The brighter the light, the stronger the current.

With hundreds of crystal wafers alternated with barriers, a sizable current can be started when the sun's rays fall on the silicon crystals.

The manufacture of silicon cells is a complicated and costly process, since the wafers must be "grown" so thin and must be carefully prepared with boron and arsenic before the cell can be made.

A silicon solar cell is about 14 percent efficient, which is higher than any other type of barrier cell. Gallium arsenide has a 10 percent efficiency; cadmium sulfide about 8; cadmium telluride, 4; indium phosphide, 3; and gallium phosphide and selenium, 1.

Using special "winding" equipment, scientists at Mobil Tyco Solar Energy Corporation have "grown" silicon ribbons up to 80 feet long. Ribbon is key component of solar cells.

MOBIL TYCO SOLAR ENERGY CORPORATION

Scientists are hard at work trying to simplify the complicated methods of manufacturing these solar/electric conversion cells.

Early solar batteries were round, but most of them now are rectangular, in standard dimensions of about 1 by 2 centimeters (.39 inch x .78 inch). A single cell weighs only 2 grams (.07 oz.) and produces about half a volt of electric power.

Connected in series, solar batteries deliver increased voltage; connected in parallel, they deliver increased current. Five cells in series can develop about two volts of power.

Several other methods are known for the manufacture of solar batteries, including thermionic conversion and photochemical conversion. One method depends on the heat differences in metals and the other in the reaction of a chemical element to light.

PHOTOCHEMICAL DIODE SOLAR COLLECTOR

One recent advance in solar cells involves photosynthesis. Photosynthesis is the process by which plants harness the energy of the sun's light in order to manufacture carbohydrates from carbon dioxide and water.

The process by which photosynthesis is accomplished chemically is not actually understood.

By simulating photosynthesis, researchers at Allied Chemical Corporation's material research center have discovered a way that might lead to a simplified method of storing electricity in a battery.

It involves the process of creating a photochemical reaction in a diode. A diode is a device like a vacuum tube. Its purpose is to cause an electric current to flow. The simple diode is composed of a cold anode and a warm cathode. When the cathode is warmed, electric current flows from cathode to anode.

In the case of the photochemical diode, the cathode is heated by exposure to sunlight. In turn, sunlight causes a process similar to photosynthesis to take place.

First of all, a diode is flooded with acidic water. A metal semiconductor sandwich is submerged in the water. When the cathode is subjected to sunshine, a chemical reaction takes place in the water that separates it into hydrogen and oxygen.

This is similar to the way water is broken down by a plant with the help of the sun's light.

Although there are many problems to be solved before the photochemical diode is ready to be put to work, it opens up a way to store incredible amounts of the sun's energy in the form of electric power.

Solar batteries that convert the sun's energy to electricity are a thing of the future, but when perfected will give the average homeowner the capability of running his entire household through the sun's energy.

For example, a solar silicon battery will enable him to produce electric power for his lights at night and to run the various small appliances in the home—electric clocks, radios, and even television sets.

In addition, electric power from the sun will operate the pumps, blowers, and sensors that control his solar heating and cooling system.

Without solar batteries, there could not be a complete solar home. Even though it is in the future, it is something that almost anyone can look forward to even now.

WHERE ARE PHOTOVOLTAIC CELLS AVAILABLE?

The only kind of photovoltaic cells available today for the homeowner are those composed of silicon, and aluminum or silver. Light stimulates the flow of electrons across the layers of material, and generates current through wires placed between them.

A photovoltaic cell can be used for any purpose requiring electricity. They can even recharge solar wristwatches.

The price of photovoltaics has dropped sharply from what it was a few years ago. It is now about $17 a watt, compared to $200 a watt. They are, obviously, extremely expensive for general use.

They can provide electricity at a tenth the cost of flashlight batteries, and they have already been put into powering machines and instruments in remote areas not supplied with electric utility power.

The problem is that it costs a great deal to make the batteries. A joint venture by Tyco Laboratories, Inc., and Mobil Oil Corporation has produced a radical new method for growing long thin ribbons of pure silicon suitable for photovoltaic panels.

Another company, Solar Energy Systems, Inc., is making cells from cadmium sulfide, a cheaper material. By the mid 1980s, it is said by some energy experts, a further fifty- or hundredfold price reduction is possible. That would put cells on everyone's roof.

Appendix

Manufacturers of Solar Components

Because of the nature of the solar equipment industry, it is impossible to present a complete up-to-date list of manufacturers in any book.

The following is, at best, a partial tally.

Principle products manufactured are listed. Symbols: S = Space Heating; W = Hot-Water Heating; C = Cooling; P = Pool Heating.

Acorn Structures, Inc., PO Box 250, Concord, Mass. 01742 (prefabricated solar houses)
Arkla Industries, 950 E. Virginia St., Evansville, Ind. 47704 (solar air-conditioning units)
Daystar Corporation, 41 Second St., Burlington, Mass. 01803 (S, W)
Edmund Scientific Co., 4632 Edscorp Bldg., Barrington, N.J. 08007 (miscellaneous solar devices)
Energex Corporation, 5115 Industrial Road, Las Vegas, Nev. 89118 (W)
Fafco Inc., 138 Jefferson Drive, Menlo Park, Cal. 90053 (P)
Fedders Corporation, Woodbridge Ave., Edison, N.J. 08817 (S, W, C)
Fred Rice Productions, 6313 Peach Ave., Van Nuys, Cal. (W)
Garden Way Laboratories, Charlotte, V. 05445 (plans for solar room)
General Electric Co., Valley Forge Space Center, PO Box 8661, Philadelphia, Pa. 19101 (S, W, C)

General Motors Corporation, 767 Fifth Avenue, New York, N.Y. 10009 (S, W, C)
Grumman Corporation, South Oyster Bay Road, Bethpage, N.Y. 11414 (S, W, C)
Honeywell Systems and Research Center, 2600 Ridgeway Parkway, Minneapolis, Minn. 55413 (S, W)
Kalwall Corporation, 88 Pine St., Manchester, N.H. 03103 (S, W)
Lennox Industries, Inc., 200 S. 12th Ave., Marshalltown, Iowa 50158 (S, W)
Libby-Owens-Ford Co., 811 Madison Ave., Toledo, Ohio 43695 (S, W, C)
Owens-Illinois, PO Box 1035, Toledo, Ohio 43666 (S, W, C)
Phelps Dodge, 300 Park Ave., New York, N.Y. 10022 (S, W, C)
Piper Hydro, Inc., 2895 La Palma, Anaheim, Cal. 92806 (S, W)
PPG Industries, Inc., 1 Gateway Center, Pittsburgh, Pa. 15222 (S, W, C)
Raypak, Inc., 3111 Agoura Road, Westlake Village, Cal. 91361 (W)
Revere Copper & Brass, Inc., 605 Third Ave., New York, N.Y. 10016 (S, W, C)
Reynolds Aluminum, PO Box 27003, Richmond, Va. 23261 (S, W, C)
Rho Sigma, 15150 Rayner St., Van Nuys, Cal. 91405 (sensors, instrumentation, solar components)
Savell System, 1645 Alston Ave., Colton, Cal. (house designs, solar)
Skytherm Process & Engineering, 2424 Wilshire Blvd., Los Angeles, Cal. 90057 (S, C)
Sola-Ray Appliances, PO Box 75, Taurt Hill, Western, Australia 6060 (W)
Solaron Corporation, 4850 Olive Street, Commerce City, Col. 80222 (S, W, C)
Solar Energy Products, Avon Lake, Ohio 44012 (S, W)
Solar Products, Inc., PO Box X-2883, San Juan, Puerto Rico 09903 (W)
Solarsystems, Inc., 507 W. Elm, Tyler, Texas 75701 (S, W, C)
Solar Systems Division (of CSI), 12400 49th St. South, Clearwater, Fla. 33520 (W)
Solar Therm Inc., Scottsdale, Ariz. (P)
Sundu Company, PO Box 351, Los Angeles, Cal. 90053 (P)
Sunsource Pacific, Inc., 501A Cooke St., Honolulu, Hawaii 96813 (W)
Sunworks, Inc., 669 Boston Post Road, Guilford, Conn. 06937 (S, W)
Tranter, Inc., 735 East Hazel St., Lansing, Mich. 48909 (absorber plates & heat exchangers)

Tyco Laboratories, 16 Hickory Dr., Waltham, Mass. 02154 (photovoltaic cells)

Westinghouse Electric Corporation, Westinghouse Bldg., Gateway Center, Pittsburgh, Pa. 15222 (heat pump)

In addition, several large directories are available containing lists of solar product manufacturers.

- *Catalog on Solar Energy Heating and Cooling Products* (ER1.11: ERDA-75)—Price, $3.80
  Suprintendent of Documents
  U.S. Government Printing Office
  Washington, D.C. 20402
- *Survey of Solar Energy Products and Services* (Y.4SC12:94-1/G)—Price, $4.60
  Same as above
- *Solar Energy Industry Directory and Buyers' Guide*—Price, $2
  Solar Energy Industries Association, Inc.
  1001 Connecticut Avenue N.W.,
  Washington, D.C. 20036
- *Solar Directory*—Price, $20
  Ann Arbor Science Publishers, Inc.
  PO Box 1425
  Ann Arbor, Mich. 48106
- *List of Solar Equipment Manufacturers.* Write:
  American Society of Heating, Refrigerating, and Air Conditioning Engineers (ASHRAE)
  (Attention: Joseph F. Cuba)
  345 East 47th St.,
  New York, N.Y. 10017

Sources of Information About Solar Energy

Further information regarding solar energy and other matters touched on in this book can be obtained by writing to the following places.

For General Information on Solar Heating and Cooling:

- The National Solar Heating and Cooling Information Center,
  PO Box 1607,
  Rockville, Md. 20850

- ERDA Technical Information Center
  PO Box 62
  Oak Ridge, Tenn. 37830

For Information about Installing Solar Systems:

- American Institute of Architects
  1735 New York Ave., N.W.,
  Washington, D.C. 20006

- The American Society of Heating, Refrigerating and Air Conditioning Engineers (ASHRAE)
  345 East 47th St.,
  New York, N.Y. 10017

- National Association of Home Builders
  15th and M Streets, N.W.,
  Washington, D.C. 20005

- Solar Energy Industries Association
  1001 Connecticut Ave., N.W.,
  Washington, D.C. 20036

For booklet "Buying Solar," guide to choosing solar equipment, write to:

- Office of Consumer Affairs, Public Affairs
  Department of Health, Education and Welfare
  330 Independence Avenue S.W.,
  Washington, D.C. 20201

For bibliography on solar energy, send stamped, self-addressed envelope to:

- ISES Bibliography
  12441 Parklawn Drive,
  Rockville, Md. 20852

Index

Italicized page numbers indicate pages with illustrations.